AGAINST ALL ODDS:
FOR THINKERS ONLY

AGAINST ALL ODDS: FOR THINKERS ONLY

BILL SMILEY

Epic Press

Belleville, Ontario, Canada

AGAINST ALL ODDS: FOR THINKERS ONLY
Copyright © 2012, Bill Smiley
Revised, 2013

All Scripture quotations, unless otherwise specified, are taken from *The Holy Bible, New Living Translation.* Copyright © 1996. Used by permission of Tyndale House Publishers, Inc., Wheaton, IL 60187. All rights reserved. Scripture quotations marked NASB are taken from the *New American Standard Bible*, copyright © The Lockman Foundation 1960, 1962, 1963, 1968, 1971, 1972, 1973. All rights reserved.

ISBN: 978-1-55452-969-8
LSI ISBN: 978-1-4600-0206-3
E-book ISBN: 978-1-55452-970-4
(E-book available on the Kindle Store, KOBO, & iBookstore)

Cataloguing data available from Library and Archives Canada

For more information, please contact:
info@forthinkersonly.com
www.forthinkersonly.com

To order additional copies, visit:
www.forthinkersonly.com
or
www.essencebookstore.com

ATTENTION CORPORATIONS, UNIVERSITIES, COLLEGES, NOT-FOR-PROFIT, AND PROFESSIONAL ORGANIZATIONS: Quantity discounts are available on bulk purchases of this book for educational, gift purposes, as premiums for increasing magazine subscriptions or renewals, or as special offers from not-for-profits for the purpose of rewarding supporters or increasing contribution income. The author is available for special presentations and speaking engagements.

Epic Press is an imprint of *Essence Publishing*. For more information, contact:
20 Hanna Court, Belleville, Ontario, Canada K8P 5J2
Phone: 1-800-238-6376 • Fax: (613) 962-3055
Email: info@essence-publishing.com
Web site: www.essence-publishing.com

Dedication

To those who are open to grasping
what others consider impossible.

Table of Contents

Acknowledgments

Thanks to my family, my friends and my medical acquaintances for your help in the many aspects of bringing this book to print. Without your collaboration this never would have happened.

Above all, special thanks to my very best friend, who tirelessly supplied me with endless fresh ideas, encouragement and provision.

About the Author

Bill approached the end of his high school years and was ready to enter university but had two conflicting academic goals: one, to pursue a career as a surgeon, and the other, to study engineering physics. He solved that conundrum by letting the outcome be settled with the single flip of a coin. Heads, it would be engineering; tails, medicine. Engineering won. Actually, medicine never really lost; it just took kind of a back seat for awhile.

Of all the things his engineering studies taught him, two stood out above the others: one, never to accept the validity of a highly questionable series of events until the evidence in their favor was so overwhelming that there could be no other logical possibility, and two, the discipline to look at complex structures and break them down into smaller entities that could be analyzed logically.

While engineering became his profession, medicine always remained a passion. Many times he would show up for a medical procedure and insist that a local anesthetic be used instead of a general and that he be propped up to see the actual operation. Then he would chat with the surgeon during the procedure, asking all kinds of questions. For most surgeons, this was a first.

Writing this book seemed to be a natural progression of the integration of these two areas of interest. Its objective would be to invalidate the authenticity of the theory of evolution, and to do it in a way never before attempted, yet so fundamentally and intuitively obvious that each and every reader would universally understand it.

Bill's engineering background taught him to look at things from a scientific perspective. Coming from a church-going family, he found, presented a challenge to his logical view of the world. "Did the world come about by evolution or by divine creation?" School taught one; church taught the other. The answer seemed obvious. Evolution has been the undisputed standard in the school systems of the world for over a hundred years. It has weathered the storms of countless challenges and yet it survives, although perhaps not as solidly as in years past.

Bill has masterfully woven his interests in medicine and logic together into a finale with conclusions that are irrefutably conclusive. For those who have wrestled with this issue, or maybe just wondered which is the real deal, by the time you've seen all the evidence you'll never again wonder. You'll know beyond the shadow of a doubt! And what makes his analysis so fresh and appealing? Bill simply presents evidence that you, the reader, are totally involved and familiar with on a daily basis: no rocks, no floods and no fossils. Then he lets you draw your own conclusions.

You will laugh, be amazed and be challenged, but most of all you will be changed.

Preface

Have you ever tried to beat the odds at something? Maybe you ran a red light, bought a lottery ticket, tried to speed without getting caught or cheated on an exam? Did you win? Did that mean the odds were in your favor, or was it just luck?

Odds entail risks, and risks evoke consequences. Look at the odds from the other side of the coin. How big would the odds need to be to totally convince you one way or the other about a given questionable situation? Hold on to your answer; you'll need it.

So if you knew how big the issue really was, would you take the chance or not? Of course that would depend on a lot of things: your depth of knowledge of the issue, whether the rewards exceed the risks, if others would be impacted by your decision or just you, if it is strictly a gamble, and so on.

You will find this book at times insightful, thought provoking, challenging and maybe even amusing. The probability of all the situations presented in this book happening at once will astound you. But the events we're going to look at did all happen at once, against all odds, and the chances of them happening as they did are so miniscule we don't have a frame of reference for them.

In these pages we're going to take a very specific focus on one of my favorite places, and yours too, I'm sure— "Home Sweet Home"—the human body. You know, your body is quite a phenomenal machine. What do you think the chances are that it evolved? And then again, what are the chances that it didn't? If you're open to some challenging evidence, read on. Even if you don't really care one way or the other, read on anyway. You'll find a lot of fascinating information with which you can astound your friends.

Look at the table of contents and you'll see that we're going to cover a lot of ground. But because there's so much to consider, we'll only look at the thinnest slice of each topic. Besides, even this thin cut should be enough to cement your opinion on evolution one way or the other.

Wait! Don't abandon the book now! You'll miss all the fun.

Much of what you will read is fundamental medical science, but you don't have to be a med school grad to understand it—like your heart's ability to pump enough blood over your lifetime to fill some twenty-seven miles (forty-three kilometers) of railway tank cars. That's a "lotta" blood!

As we go through the following sections I'll be asking if you think a particular characteristic of your body could have happened by chance or not. Just to pick a starting number, let's think the odds for each happening would be one chance in a million. A million doesn't seem like all that much these days, so if you're comfortable with my arbitrary choice of chance, great, but if you'd rather choose a different number, be my guest.

Other circumstances will be intuitively obvious but challenging, like how easy it would be to survive if the only change random chance made to your body was to have your head attached facing backwards. Build a mental picture of it and try a few basic motions, like giving your sweetie a hug and kiss, or fishing, or making a sandwich. Why, you wouldn't know if you were coming or going! (I know that was a pun, but my wife made me do it.) Survival would indeed be doubtful.

I make one caveat right up front: I will not be drawing any personal conclusions in what follows, just making observations and the occasional musing. Conclusions will be your job. Much of what you read you will know experientially to be true or it can be verified in any number of ways. I will present the facts. You tally the possibilities as you see them and check what the collective numbers look like at the end.

That sets the stage; we're ready to begin. So sit back, learn some new stuff about how you tick, have a laugh or two, and enjoy the possibilities.

Against All Odds

We've all heard the story, or its variations...Way back when, millions or billions of years ago, the skies were dark and heavy with rain. Lightning flashed from horizon to horizon. Thunder reverberated menacingly across the sky. The earth was impassable, a gooey soup. Then, suddenly, and with energy beyond comprehension, a humongous lightning bolt struck the ground, vaporized the goop, changed its molecular structure, and "Voilà!" the beginnings of what would become human life. That story has been taught and told so many times that to challenge it would seem almost blasphemous. Besides, why would anyone want to try when there are such adamant proponents on each side? But that's exactly what we're going to do. We, you and I, are going to end this stalemate once and for all. And our human bodies will be our proving grounds.

We take our bodies so much for granted, often to the point of becoming blasé about the simplest of things. For example, we don't need to keep reminding our hearts to pump and our lungs to breathe because all that somehow happens without us ever having to think about it. When we get a cut we know that in a week or so it will all be healed. Children are born. Colds and flu come and go. So what? Well, how does the body do all this? These are intriguing considerations, wouldn't you agree? As we explore the wonders of how we are put together and operate, you will gain a deeper appreciation of the intricacy of this structure we call our "body."

Regardless of how we came to be, within the woman's egg (a single cell about the size of the period at the end of this sentence) and the man's sperm (1/600 inch or 0.05 millimeter long) are all the genetic data needed to reproduce. That's not news to any of us, but consider what comes from that union. At one level we can rhyme off the body's organs and structures. But what about the way these structures grew from a single cell and became attached: how do tendons link muscles to bones, and how do discs form between vertebrae? What about the sequence of development: for example, do the kidneys, brain and heart all form in the same time frame, or does one wait until another is more or less developed?

As I write this book, my next-door neighbor is renovating a small bungalow into what I would call a monster house. It's interesting to watch the sequence in which things happen. First the basement is excavated, and then forms go in place and are filled with concrete. Next, the main floor goes in, walls are framed, the plumbing goes in and the electrical wiring is completed. Finally, the drywall goes up, baseboards are put in place, and the interior is painted; and that's more or less how the sequence happens. Clearly, if the construction occurs in the wrong order there are going to be major issues. If the walls were finished and painted before the pipes and wiring were in place, the finished house might look finished, but it wouldn't be at all functional.

If it were up to me to specify the sequence in which the body developed in the womb, I couldn't even begin to comprehend the complexity. What would happen, for example, if the skeletal system developed in the womb with an adult-sized head or with bones that were not pliable? That's a slightly easier question, I think. The mother would die trying to deliver, and likely as not so would the child. Was this the case "in the beginning"? If it were, we just might not be here. Let's find out.

Your body consists of several completely unique systems, including circulatory, nervous, reproductive, skeletal, muscular, digestive, immune, respiratory, glandular, hormonal, and a few other odds and sods. System by system, we'll consider the chances that from the very beginning things were the way they are. We'll start by looking at the body's fundamental building block, the cell.

Building Blocks

Cells: Your Body's Building Blocks

Have you ever thought you knew exactly why you were doing some particular thing only to find out later that you were doing it for the wrong reasons?

My wife and I had been like that with vitamin and mineral supplements. We reasoned that if we took a quality brand we would stay healthy. Then one day my wife was sitting beside a friend at a conference. Her friend was in a thin blouse while my wife had on a blouse, a sweater and a jacket, and she was still cold. Her friend matter-of-factly mentioned that my wife needed vitamin E. She was right! Vitamin E cured the chills almost immediately. That experience led us on a search for a better understanding of how our bodies work. After all, we want to be in total health until the day we die. Who doesn't, right?

In our research, we discovered something so fundamental it embarrassed us that we hadn't figured it out a long time ago. We had been taking supplements for specific reasons: one to promote bone development, another for improved heart function, etc. Yet we never considered that the body is built of cells and that if we didn't provide the right nutrition to the cell walls, the higher order systems of the body that those cells make up would never function at their best.

Here's how it works. You go out for a wonderful meal. It gets digested. Your body retains nutrients, and the rest is eliminated. It's the same at the cellular level. Each of your cells has a membrane that surrounds it, and with a proper diet that cell wall will permit nutrients to enter the cell and wastes to be eliminated. But, if that membrane is not getting the nutrients it requires, very little value is going to get into the cell, and not enough of the waste is going to get out. The result is that we are not going to be as well as we could be.

It intrigues me that this membrane is there with this nutritional gatekeeper function. What if the cell membrane didn't enable both nourishment in and waste out? (That, by the way, was our first one in a million question.)

SELF-REPLICATING

DNA, deoxyribonucleic acid, is a self-replicating material found in nearly every living organism. As complex as DNA is, there remains a question fundamental to life itself: when and how did DNA acquire its ability to divide? (That was the second one in a million question. From here on, you keep track.) This is a crucial consideration, since without division there would be no growth, and without growth, no life form would ever have existed beyond the single cell.

BUILDING

Your body is an integrated symphony of some 73 to 100 trillion cells (100,000,000,000,000), and each one is supplied by your body's pipeline of life, the bloodstream. Have you ever wondered what holds all our cells together? Science calls it "laminin," and it's kind of a genetic Crazy Glue. Actually, it's a protein that helps bind molecules together. Its four arms can bind to four other molecules as well as to other laminin molecules. This is the capability that helps anchor the body's cells into meaningful structures (organs) and into their surroundings. So technology has told us what this cellular glue is, but as of yet there is no satisfactory answer to the question of when it appeared on the scene. Without this glue our cells would have had no connectivity with each other, and all could have drifted off here and there. How then could there have been any kind of life?

REPAIRING

When I was a young teenager, while trying to impress a girl I got my hand caught in the rapidly rotating spokes of a bike. Dumb? No kidding! It was just the friction of my hand against the spokes that brought the bike to a stop. Painful! The back of my hand was black and blue and red and white and an inch thicker than it had been ten minutes earlier. My mom freaked, the doctor shook his head, and I sobbed. Today, as I type this section, there are no scars whatsoever, and I don't have to wear an oversized mitt on that hand. Why? It's because our bodies somehow have this marvelous ability to heal and repair themselves. But if cells had not been self-repairing from day one, how long would life have lasted?

REPLACING

Early in a child's life, every part of the body, with the exception of the brain and nervous system, is replaced at least once a year. The body's cells are constantly dying, being disposed of and being renewed. If this had not been the case from the very beginning, life would have died off with the death of the first cell. But we know that's not the case. What are the chances?

OK, now that we've got a bit of a perspective on cells, let's explore the various structures they make up, starting with the circulatory system.

Round and Round We Go

The Circulatory System

As part of my annual physical this year I had an echocardiogram. After the nurse completed the eighty or so images she needed, I explained to her that I had a huge interest in medical techniques and asked her if she would explain the process to me and show me some of the images she had captured. To my great delight, she was a talker and started to bring up the images. Only they weren't just images, they were videos too, complete with the sounds the blood makes as it swishes in and out of the heart's four chambers. Fascinating! I came away from that session very impressed by this little fist-sized organ that is powerful beyond what we ever take time to consider. It is the most remarkable of all the organs in our bodies. Consider the following.

THE HEART CYCLE

If you make a fist and twist it around so you can see it from all sides, that's about what your heart looks like. It's actually quite small and weighs about ten ounces (0.3 kilogram). This muscle beats about seventy times a minute in an adult as it circulates blood throughout your body. Every one of your body's seventy-three trillion (73,000,000,000,000) cells needs blood to nourish it and to eliminate its wastes; if that couldn't happen, the cells would die rather quickly, and so would you. As you read on, consider this question: could this amazing organ have gradually evolved, or would it have had to be fully functional from the outset of humankind?

Your heart operates in a repetitive and very precise sequence of steps. To start with, it circulates blood in a one-way path throughout your body. Its four chambers each squeeze the blood forward as the heart muscles contract and then relax. As the blood moves from chamber to chamber, heart valves, which are thin pieces of muscle not much thicker than a sheet of paper, prevent the blood from flowing the wrong way. If we follow the blood flow, we'll see just how intricate it is.

To begin with, blood that has traveled through the body and is now depleted of its oxygen flows into the right atrium chamber of the heart. From there it flows to the right ventricle, which in turn pumps it to the lungs, where waste gases, mostly carbon dioxide, are exchanged for a healthy supply of oxygen. The oxygen-rich blood then flows from the lungs into the left atrium, then to the left ventricle and then on to all parts of the body and ultimately to every one of those seventy-three trillion cells. And the cycle repeats.

YOUR CIRCULATORY SYSTEM'S ROAD MAP

When I bought an in-ground pool some years ago, I had to decide how often I wanted to have the water circulated through the filter, and the answer to that question determined the size of the pump I would need. Consider the heart pump. Your body has about 5 quarts (4½ to 5 liters) of blood that must be circulated through approximately 60,000 miles (96,500 kilometers) of veins, arteries and capillaries. That's enough blood vessels to circle the earth two and a half times.

Imagine the engineering challenge there would be to design a ten ounce (0.3 kilogram) pump—the weight of a human heart—that would improve on this feat: your body's blood cells make a complete round trip from the heart to your extremities and back to the heart again, once every sixty seconds. As if that's not challenging enough, consider what this tiny pump will have to handle over the average seventy-five-year lifespan. The heart pumps about 1.3 gallons (5 liters) a minute; that's 1,900 gallons (7,200 liters) a day. Over the course of a seventy-five year lifetime, that's 51 million gallons (197 million liters), enough blood to fill a string of railway tank cars 27 miles (43 kilometers) long.[1]

Stop for just a minute and reread these last few paragraphs. They describe something absolutely staggering when you think about it.

The very first time I saw the woman who would someday become my wife, I think my heart literally skipped a beat. That was OK; I had lots more beats and could afford to skip one. You see, if you were to count your number of heartbeats per day they would number around 100,000.[2] During your lifetime, your heart will beat continuously some 2.4 billion times.[3] Once it starts, it doesn't stop until you die. The only time it ever gets to rest or repair itself is between beats, and even when your body is at complete rest, your heart is still working twice as hard as the leg muscles of an average adult male while sprinting. Putting the results from these last two paragraphs together, what are the chances that this is a "Mother Nature" special?

HEART HARMONY

Medical science has made many intriguing heart discoveries, but consider just this one. Every heart cell has the ability to create an electrical current that signals itself to contract; this is called cardiac contraction. If we take a laboratory look at these cells we see an inexplicable pattern. Set these cells completely apart, and each will beat with its own frequency, but let them touch again and after a few days they will all begin to beat with a frequency equal to that of every other heart cell. In other words, they will all beat in unison.[4] Consider that if this were not the case, the result of these random firings would be chaotic to the beating of the heart; in other words, there would be no heartbeat and, hence, no life. So then, would you consider the odds of this strange but essential cellular beating to be one in a million?

And what about the intricate design of the heart and its connectivity to the lungs to allow your blood to give off waste gases and pick up essential oxygen to nourish the cells along your 60,000 miles (96,000 kilometers) of capillaries; would you rate those odds as one in a million?

And finally, the odds of all of this happening concurrently and instantly from day one so the blood nourishment to every cell of the body is never in jeopardy, is that another one in a million?

You Gotta Have Nerve

The Nervous System

Of all the parts of the human body, the nervous system is by far the most elaborate and complex. Let me compare its operation to that of your home computer. Like me, you may work with a word processing program called Microsoft Word. Say you have a file called *Résumé* that you want to modify. When you click "Open *Résumé*" and then press "Enter," that causes the computer to interrupt what it's doing and to pass control to a software routine in the Windows operating system that resides in the computer's central processing unit (CPU). Windows will search its directories to find where *Résumé* resides on your disk, and then it will send a signal to the disk drive to tell it that it has information to be transferred to your Word session. While the operating system waits for the disk drive to respond to this request, it goes about other business. When the disk is ready to transfer your *Résumé* data back to Word, it causes whatever is currently running in the computer to be interrupted again so the contents of *Résumé* can be sent to Word.

This complicated process of interruptions continues until you end up with your *Résumé* being displayed on your screen. And so it is with the nervous system, only with a lot more complexity.

When we talk about the nervous system we are collectively referring to two systems: the central system and the peripheral system, two distinct parts that must act in perfect harmony and coordination with each other for your body to function properly. It's a gigantic electrochemical computer and a complex network of communication channels made up of billions of nerve cells that link to every part of your body.

When organs like your skin, ears, nose and tongue react to external stimuli, they send signals to your brain and spinal cord (your central nervous system) by means of sensory nerves. Your brain and spinal cord interpret these signals, decide what your body needs to do in response, and then send the appropriate impulses via motor nerves to the various glands, muscles or effector organs. For example, something hits your eye, and your eyelid automatically shuts as a precautionary measure. Without this incredible system of nerves, you couldn't hold this book, or read it, or understand what it says; nor could you walk, talk, or perform any other bodily function. In short, you could not be.

THE CENTRAL NERVOUS SYSTEM

Your brain, spinal cord and retina collectively make up your central nervous system. As with the CPU being the "brains" of the computer, your central nervous system is the central processor for your entire nervous system. It, in turn, connects to the peripheral nervous system.

Your brain, of course, is housed and protected by your skull, and your spinal cord is encased in a bony channel that runs through each spinal vertebra from top to bottom. Consider the skull first. I am not blessed athletically, and I will never again put on ice skates, because my head has this biased propensity for unsavory encounters with the ice! After being knocked out cold three times, I have come to the irrevocable conclusion that without a skull I would no longer be alive.

We need to consider this bony channel. At each vertebra, nerves branch out to your limbs and organs at that particular level in your body, at one level your heart, at another your lungs, at a third your kidneys, and so on. This is how your body links your brain to the rest of your body. There's a stroke of genius here that most of us take almost totally for granted, and that is the body armor that protects the brain and spinal cord. When did these protective bony shields come about? Think about the gravity of even a minor back or head injury without that channel. If the protective shields had not been there from the very beginning, it wouldn't have taken very many bumps to seriously compromise progression of the human race. What do you think; chance or design?

BRAIN FUNCTIONS

If you could look at the inside of your computer's CPU (its "brains") you would see that there are various parts that control specific tasks correctly, such as arithmetic calculations, logic, input and output, control, interrupts, keyboard, displays, and so on. In this respect, your brain is quite similar in that it has various sections that deal with specific requirements and that communicate with each other through a specialized network of nerve cells. This compartmentalization is truly a wonder. For example:

- One area manages speech, personality, emotions, memory, intelligence and your ability to feel.
- Another regulates pulse, appetite and your body's automatic processes.
- A third regulates your metabolism, growth, stress response and sexual maturity.
- Yet another receives and analyzes signals collected by your sensory nerves and then transmits them on to other sections of your brain, where they are further analyzed.
- The center of your brain is your brain's information gatekeeper; it manages the flow of all impulses flowing into and out of your brain, and the rear part of your brain manages movement, balance, and automatic functions like blood pressure, heart rate, digestion and breathing.

Your brain contains around one hundred billion (100,000,000,000) cells called "neurons"—cells that use electrochemical stimuli to process and transmit information from various parts of your body via your central nervous system. If that's not impressive enough, consider that there are even several special types of neurons:

- Motor neurons process signals from your brain and send them to your muscles for contraction and release; they are critical to all your bodily motion.
- Sensory neurons respond to several stimuli, including the senses of smell, taste, touch, hearing and sight in your sensory organs.
- Interneurons do exactly as their name implies: they interconnect neurons within your brain and spinal cord.

Astonishingly, each neuron is linked with up to 10,000 synaptic connections, a "synapse" being the junction between two nerve cells, consisting of a minute gap across which electrical impulses can pass. When you consider the size of an adult brain, weighing around 3 pounds (1.4 kilograms) and occupying less than 1 quart (2 liters) in volume, the points of connection are a staggering 100 to 500 trillion synapses: 100,000,000,000,000 to 500,000,000,000,000.

Was it happenstance that this complex structure developed purely by chance? If the processes were evolutionary, how would the early life forms ever have survived right from the get-go without the myriad functions managed by the human brain?

THE PERIPHERAL NERVOUS SYSTEM

The principal function of your peripheral nervous system is to connect your limbs and your twenty-two internal organs to your central nervous system. It is made up of all your body's sensory organs and nerve cells that are not found in your brain or spinal cord (i.e., your central nervous system). When nerve cells respond to stimuli, from either inside your body or something outside, those responses are detected and collected by your peripheral nervous system and are then sent on to your central nervous system for interpretation and action.

The medical community differentiates two parts to the peripheral nervous system, and they assign complicated names to each: the sensory-somatic nervous system, and the autonomic nervous system. If I were a professor, which I'm not, I'd tell you to memorize what follows and be ready for a test on Thursday, but that's not the level I want to take you to. Rather, I simply want to impress on you the astonishing organization, structure and functionality of these two subsystems.

SENSORY-SOMATIC NERVOUS SYSTEM

First, let's consider those bodily functions managed by your sensory-somatic nervous system. It controls taste, tongue muscles, balance, hearing, smell, facial muscles, vision, eyelid and eyeball muscles, salivary glands, swallowing, and head and shoulder movement. If it weren't for this part of your peripheral nervous system, you would have no conscious awareness of things happening outside your body—your external environment; nor would your body have any way to control motor activities to cope with it.

AUTONOMIC NERVOUS SYSTEM

Your autonomic nervous system monitors conditions within your body—your internal environment—and brings about appropriate changes in them where necessary. Its actions are largely involuntary; they happen without your consideration or intervention, so you never need to think about them. Consisting of nerve cells that run between your central nervous system and various internal organs, your autonomic nervous system performs numerous functions:

- It manages all functions you don't consciously control, including breathing, heartbeat and blood pressure.
- It controls all of your physiological responses to pending danger and the "rush" you feel when there's a sudden release of adrenaline into your bloodstream.
- It manages your brain's "always on" status, digestion, waste elimination, sleep and relaxation

management and its refreshing effect on your body.

- When strong emotions like anxiety, anger, hate and fear overpower you, it acts like an emergency control system and adjusts many of your internal organs to cope properly with the emotional changes.
- Whenever your body enters into a state of stress, your autonomic nervous system causes things to change so your body can adapt to the stress, such as, your heartbeat increases, blood vessels constrict, glands secrete more, and salivary and digestive glands secrete less.

To get a deeper perspective on the sophistication of the workings of your body, ponder any one of these functions and ask yourself if the intricacies of your nervous system are not way outside your boundaries of comprehension. How did all this inconceivable control system come to be? By accident? By evolution? By plan?

You don't have to have definitive answers to these questions, but there is still an unavoidable uncertainty that demands an unambiguous answer. What do you think the likelihood would be that humanity could have ever gotten off the ground without the complete capabilities of our nervous system being fully in place from humanity's origins?

Multiplication Matters

The Reproductive System

IN THE BEGINNING

In all my years of reading medical books, watching documentaries and teaching programs on reproduction, and just being plain inquisitive about nature, I have yet to see any evidence that there has ever been any kind of fertilization from a same-sex partner. If that big ol' lightning bolt of long ago did, in fact, create a self-fertilizing organism, where is the evidence? And if that were the case, what caused all of humankind to change from self-fertilizing to what we see today, cross fertilizing between man and woman? And why? What was the impetus for such a change? What are the chances that male and female (of any species for that matter) were created sexually compatible and at exactly the same time? Think about it: if this were not the case, then there would be no reproduction and you wouldn't be reading this book.

SEXUAL ATTRACTION

Regardless of all the physical aspects of reproduction, there is another consideration that cannot be omitted. You guessed it: hormones. Suppose all the various systems of the body did come about totally by chance but without that powerful hormonal force—the drive to procreate. The chances of the human race ever getting off the drawing board would have been in serious jeopardy. Wouldn't you agree? So what are the odds of this essential hormonal drive being there from the very beginning?

SPERM

In today's information age, it is unlikely that you, the reader, have not heard about the male gland called the prostate. But how many of us know how totally essential this little gland is to human life? You see, sperm can't exist in an acidic environment and during ejaculation it will leave the man's testicles, enter his urethra (that duct that conveys both his urine and semen through the penis) and journey on into the woman's vagina. Herein lie two major reproductive obstacles. First, any urine remaining within the man's urethra is acidic and will kill off his sperm as it enters from the testicles. And if that's not bad enough, the woman's reproductive tract is also acidic. So if the sperm doesn't die within the male, it will die when it enters the woman. That certainly sounds bleak, doesn't it! Enter the prostate gland.

The prostate manufactures an alkaline fluid that is secreted into the male's urethra "just ahead" of the sperm, thereby neutralizing the acidic urine; and, of course, that same alkaline fluid is first to enter the vagina to neutralize its acidic fluids. This creates an "all clear" for the sperm to successfully find and fertilize the woman's egg. Yahoo for the human race!

But wait! Think about it! Is it plausible that this is another one of those so-called coincidences that just happened to be there from the very beginning? What are the odds?

Even with its neutralized passage, the sperm's challenge isn't over. In each release, the sperm count can be numbered in the millions, but many die off in their journey or never find the woman's egg. If only a few sperm were produced for release each time, the chances of reproduction and life would be seriously reduced. If this is an evolutionary discovery, how did our life form ever figure out the need for huge releases of sperm, let alone produce them in time to prevent an embarrassing and untimely extinction?

PHYSICAL SEXUAL COMPATIBILITY

Have you ever wondered how it is that the two sexes so perfectly complement each other sexually? We each have the same basic appearance front to back (that's handy); our arms and legs are similarly located; one doesn't have to stand on his head to kiss his mate. You get the picture.

At the risk of seeming crass, let me pose another consideration. When did it come to be that the external male and female reproduction "plumbing" was 100 percent compatible—one size fits all, if you will?

BIRTH CANAL

Whenever I answer a phone call from my mother-in-law, there's always a short conversation with me, after which she asks, "Is my daughter around?" to which I reply, "In all the right places." She chuckles, sometimes for real, sometimes just politely. Guys, we don't have curves—well, not like a woman does, and that is a good thing. However, you women have hips like no man has hips. Your pelvis is uniquely shaped to permit the passage of a baby from your womb at birth. If this pelvic channel had not been large enough, babies would not have been able to pass through the birth canal. Then what do you suppose would have happened to the human race?

Umbilical Cord

Within the umbilical cord, which connects the placenta to the developing fetus, one normally finds one umbilical artery and two umbilical veins. The function of the umbilical artery is to supply the fetus with nutrient-rich oxygenated blood from the mother's placenta, and the umbilical veins return the oxygen- and nutrient-depleted blood from the placenta back to the mother's heart. The cord itself is of sufficient length to allow the child to move about in the womb with some freedom. Curious, isn't it, that the cord's artery and veins would connect to just the right places within the developing child to nurture and sustain life during its nine month journey to birth. Without this replenishment system from the very beginning, is it likely that human life could ever have been?

Placenta

Of all the parts of the body, the placenta is probably the most disregarded and unappreciated; but if we could apply a title to it, "Selfless Servant" would be an apt fit, as we will soon see. The extraordinary importance and complexities of this organ are totally essential to birth, yet after the baby is delivered, it is literally disposed of and forgotten.

We couldn't get more fundamental than a single sperm fertilizing a single egg and a new life beginning. That alone defies words. Why, just trying to understand the genetics and chemistries involved is staggering! But wait; within hours that single cell will divide and form two cells. Recent studies have discovered that one of these cells is destined to be the baby, and the other is destined to be the placenta.

Within the first few days after fertilization, the placenta will begin manufacturing hormones that will prepare the lining of the uterus to receive the embryo's attachment, called "implantation." Over the next month or so, more placental hormones will begin to adjust the mother's physiology in a number of ways that will ensure her baby's growth.

As the child develops in the womb, the only organ that is not supported by the placenta is the infant's heart. The placenta serves to support the baby's other critical organs until the child is ready to enter the world and fully function on its own. Here are some of them:

- As the lungs, the placenta takes carbon dioxide from the baby's bloodstream and replaces it with oxygenated blood from the mother's blood.
- As the liver, the placenta transfers important nutrients from the mother's blood cells to the baby's bloodstream.
- As the digestive system, enzymes within the placenta digest intestinal waste from the baby.
- As surrogate glands, the placenta produces a number of hormones necessary to the development and well-being of both the baby and the mother.
- As the immune system, the placenta manufactures various ingredients that prevent infection.
- As the kidneys, the placenta takes urea—the main component of urine—from the baby's body via

the umbilical cord and passes it to the mother's bloodstream, where it is then processed and expelled via her kidneys and renal system.

What would the chances of human life be if all these functions had not been in place and fully functional from the very beginning? One in a million?

THE EXPULSION OF THE PLACENTA AFTER BIRTH

While the baby is still in the womb, the placenta is firmly attached to the mucous lining of the uterus. At birth, the baby enters the world, still attached to the placenta by the umbilical cord. Once the cord is severed, the uterus will begin to contract to expel the placenta, which at this point serves no additional purpose. With all its fantastic functions, it becomes a throwaway organ.

As the placenta is expelled from the wall of the uterus, some of the uterine wall is severed with it, torn off, if you will. That tearing ruptures close to twenty large uterine arteries, resulting in a hemorrhage that could produce an immediate blood loss of close to 1 pint (0.5 liter) per minute if there were nothing to stop it. If completely unchecked, all of mom's blood would be lost within the short span of about ten minutes. Compound this with another high-risk condition: during pregnancy the ability of the placenta and uterine blood vessels to clot blood is suppressed. With the expulsion of the placenta there suddenly exists a situation analogous to a hemophiliac with twenty-odd blood vessels rupturing all at once. Why is it then that we don't see more mothers hemorrhaging to death during childbirth? More precisely, how did we ever develop as a race of people if this situation went unchecked from the very beginning of human life?

The answer is truly astounding, and it has to do with sphincter muscles, round muscles that surround and serve to close off an opening or a tube, such as the anus or the stomach as it leads into the intestines.

Consider the placenta as it comes loose from the uterine wall; at the root of each detaching artery there is a small sphincter muscle that tightens as the artery is severed, immediately stanching the flow of blood from the wound. Total blood loss: about 1 pint (0.5 liter). Awesome or what? What are the chances this was an evolutionary development?

THE HOME STRETCH

Here's an intriguing little fact you can file in the for-what-it's-worth category. At delivery, the uterus can be up to five hundred times its normal size, and following delivery it will start to return to (close to) its normal size. What about the ramifications if this were not so?

NEWBORNS

When my fourth granddaughter arrived, it was a tough delivery. Natural childbirth was not progressing as one would wish; major complications had set in, and at the very last minute an emergency caesarian section was scheduled. My daughter-in-law was not well, and my son was a puddle of emotions. An hour after the delivery, my son and I went to the neonatal intensive care unit to see the new addition to our family. She had wires and tubes everywhere. Oxygen lines and IV drips were running, and her head was completely covered with some kind of fabric mask that held a number of the tubes and sensors in place. Her little chest was heaving with a rhythm that looked like nothing I'd ever seen before; it was very scary. I marveled at the equipment and the technology that were in place and the skill and calm understanding of the nursing staff. How on earth, I wondered, would she or her mother have ever survived without all this help at their disposal? That same question would also apply to those at the beginning of the evolutionary timeline. Is the answer self-evident?

NURSING

We take a lot of early childhood nourishment for granted these days. Store shelves are full of a broad variety of infant formulas—take your pick. But what if the earliest mother did not have breasts and a source of human milk with which to feed her child? How long would the child have survived, and on what? Without crucial nourishment, among other essentials, the infant would have been an easy target for disease, another serious compromise for the beginnings of the human species.

MOTHER'S SKIN

Skin is amazing, the way it can expand and contract like elastic. Consider the implications during pregnancy. As the fetus develops, the mother's skin stretches to accommodate her growing child within. Her skin is also strong enough to prevent the baby from bursting forth through her abdominal wall. And after the birth, with a little exercise and appropriate diet, mom can lose the extra weight and look as good as she did before her pregnancy. That's fascinating!

But what if mom's skin didn't stretch? Something would have to give; either the child would be deformed and die within, or the mom would split open and likely die, too. Either way, humanity would not get much of a chance at further development. Is this a thumb up or a thumb down for evolution?

GROWTH

If you have kids of your own, you know how parents are always marking dates on a wall somewhere to show their kids' height and weight as they grow. Yet, isn't it intriguing how their bodies mature in a perfect proportion to all their individual parts? Throughout life an extraordinary symbiosis exists between muscles, tendons, ligaments, bones, etc., as they all develop in exacting proportion to each other, ensuring the first steps the child takes and the millions that follow. And in the midst of these seemingly natural changes, a great many unseen events are taking place.

Today, for example:

- Your bone marrow will produce approximately 2,500,000 new red blood cells every second to replace those no longer useful to your body.
- Twenty-four billion body cells will be replaced.
- Dead cells will ultimately go to your kidneys for excretion.
- Your kidneys will have 400 gallons (1,520 liters) of recycled blood pumped through them.
- You will perspire between 1 to 2 pints (0.5 to 1 liter).
- Your 2,000,000 skin pores will efficiently cool your body by removing excess heat via the evaporation of sweat.
- Of the 88 pounds (40 kilograms) of oxygen you breathe, 20 percent will go to your brain.
- You will breathe 23,000 times.
- You will inhale about 1 pint (0.5 liter) of air 14 times a minute.

- Your lungs will move about 353 cubic feet (10,000 liters) of air.
- If today you are under stress, your blood and air volumes will increase tenfold.
- Your heart will beat 100,000 times.
- Your eyes will blink over 27,000 times (10,000,000 times in a year).
- Your brain will lose about 7,000 brain cells if you are over thirty-five, never to be replaced.
- One-quarter of your brain will control your eyes and what you see.
- Your brain will be continuously active and will think more at night than during the day.
- Your scalp will produce, on average, over 100 feet (30 meters) of hair fiber.
- Men will lose about forty hairs today; women, seventy.
- Your mouth will produce about 3 pints (1.5 liters) of saliva.
- The ridges in your hands and fingers will help you to hold things.

This week:

- Half your red blood cells will be replaced.

This month:

- Your fingernails will grow 1/8 inch (0.32 centimeters);
- You will shed the entire outer layer of your skin.

This year:

- You will shed 10 billion skin flakes; that's 4½ pounds (2 kilograms).
- Each eyelash will be replaced twice (average lifespan is five months).
- Your ears will each grow about 1/100 inch (0.25 millimeters).

Dem Bones, Dem Bones

Cells: Your Body's Building Blocks

THE SKELETAL SYSTEM

Fundamentally, our skeletal framework has two purposes: one is to protect our internal body parts, and the other is to give us shape and mobility. It's no revelation to any of us that life would be drastically different without it, but let's take a look at just how intricate and well-purposed this system really is.

BODY SYMMETRY

Whenever you've looked at a picture of a skeleton, or perhaps seen an actual (or replica) skeleton in your doctor's or chiropractor's office, have you noticed that it is symmetrical? That symmetry is not just the same bone copied, moved to the other side of the body, attached and, Voilà! It's actually a mirror symmetry; you could not take a bone from one side of the body and use it as a substitute on the other side. Intriguing! Don't you just have to wonder how that developed, be it by design or by accident?

BONE TYPES

You were born with about three hundred bones,[5] but by the time you reached adulthood, many had fused together, and you then had two hundred and six.

Consider what the various bones look like and you will see an interesting alignment of shape to purpose. Generally speaking, there are four types of bones, each with its own specific set of responsibilities:

- Flat and circular: Like the vertebrae found in your spine.
- Long and circular, strong and flexible: Like your ribs, protecting your internal organs.
- Long, thin and light: Like those in your limbs.
- Flat, strong and light: Like those in your skull and hips.

YOUR BACKBONE

Visualize for a moment watching a world-class gymnast competing for Olympic gold. Picture, too, that you are wearing glasses that let you see only her skeleton as her floor mat routine is in progress. You watch the extraordinary flexibility and mobility of the human back and its ability to support an unbelievable variety of running, bending, turning and twisting motions. We need to take a closer look at this wonderful "machine" we take so much for granted.

Altogether there are thirty-three irregularly shaped bones in your back called vertebrae. Each vertebra is cushioned from its neighbor by tough cartilaginous discs that act as shock absorbers. Within each vertebra there is a channel that safely encases your spinal cord, the information highway between your brain and every part of your body. Working down from your head,

- Your cervical vertebrae support your neck and head and all the motions your head is capable of.
- Your thoracic vertebrae are next, and they support your ribs.
- Your strong lumbar vertebrae support the majority of your upper body weight.
- The back wall of your pelvis is made up of your sacrum's five fused vertebrae.
- And finally, your coccyx (whoever invented that word?), often called the tailbone, is made up of four fused vertebrae.

Your vertebrae are held together and stabilized by tough, flexible fibrous tissues called ligaments. In fact, your entire body contains intricately linked networks of bones and moveable joints, which are cushioned by self-lubricating surfaces that allow you to move flexibly and painlessly for a lifetime. Can you appreciate that without muscles, ligaments, tendons and nerves from the very beginning, motion, bodily shape and structure would have never been possible and life as we know it would never have started?

Because many of you likely have experienced some degree of back pain, or know those who do, here's something worthy of consideration: without these vertebral discs as part of our backbone from the very onset, early mankind, whether you believe he started out on two legs or on all four, would not have been able to run to catch food. The back pain of vertebra grinding upon vertebra would have been too much to bear. The human race would have died out before it ever started. Does it seem plausible, then, that these discs (cushions) have been with us from the very beginning? What does that suggest to you: happenstance, or design?

YOUR HEAD

Your skull is made up of fourteen facial bones and eight flat cranial bones, those that encase the brain. Without this hard, bony protective cage, your brain would be irreparably damaged from bumps and knocks. About just your cranial bones: at birth they are supple and can easily deform during the birthing process, yet when you approach two years of age they will begin to fuse together. Consider how critical that is to human existence: without that flex, it would be difficult, if not impossible, for the baby to pass through the mother's narrow birth canal. What are the chances this was the case from the very beginning?

Here's a little exercise. Yawn! If that was a real yawn, put the book down and continue later. If you are still wide awake, you've just moved the only bone in your head that has freely movable joints—your jawbone. But what if it were not free to move? There would be no access for food or fluids to enter your mouth. How would you bite, tear or chew? How would you keep food in your mouth? How would you survive? Was the jawbone always thus?

Let's change the tone and look at something on the silly side, yet not outside the realm of possibility. Picture our earliest human ancestors exactly as we are today, with one slight modification: their head looks backwards, not forward as ours does. What would life have been like? How easy would it have been to flee danger, chase game, fish, prepare food, build things, or any of the thousands of other things we can do with our heads attached as they are? Would life have been possible?

SHOULDERS, ELBOWS, KNEES

If that last set of questions set you thinking, here are some more. Build a mental image, kind of like a cartoon, and think about the ramifications:

- What if just our elbows hinged backwards?
- What if just our knees hinged outward?
- What if both our feet faced sideways, to the left?

Just days after I wrote these three questions, I happened to be tuned in to a program that was focused on the medical challenges faced by many Africans. To my total amazement, they played a video clip of just such a man. From birth, his feet pointed backwards and his knees hinged backwards. He was the unfortunate epitome of my preceding posits. His only mode of movement was to crawl on hands and knees. How very sad! Then I started to have a much different appreciation for how we are formed and move.

PUTTING IT ALL TOGETHER

This marvelous structure, our skeletal system, is a living, growing structure, and we never even need to think about it. It just happens. And when instructed by our nervous system, it works in sublime harmony to produce and control our every movement. An accident of nature? What do you think?

Movers and Shakers

The Muscular System

There are three areas we will look at: muscles, tendons and ligaments. Without any one of these, life would not be. Starting with the muscles, let's see what surprising insights we can find.

MUSCLES

As a kid I pretty much took muscles for granted, until that first night I joined a Judo class. It was the year I graduated from university, and three of us decided we needed some regimented exercise in our lives. We enrolled and showed up for night classes, not knowing that the beginners' group we were assigned to was already in its fourth week. Our sensei—teacher, that is—took us aside and told us to lie on our backs, put hands in our belts and chins on our chests, and hold it. We did. He disappeared.

At first it was easy, but then it got harder. I didn't know at the time that an adult's head weighs about 15 to 20 pounds (7 to 9 kilograms). After five minutes it felt like a lead weight. My chest hurt, and I was starting to sweat a lot, but there was no way I was going to be the first to drop my head back to the mat; it was a guy thing, a combination of testosterone and stupidity. Stupidity was winning out—actually not just for me, but for all three of us.

By the time the sensei returned we were all puddles. Fifteen minutes had passed. He looked at us, gave us a silly look that made me want to kill him, and told us to get up and join the rest of the class. Later that night we learned that that exercise never goes for more than five minutes. Nice!

The next morning when I got out of bed, my chest muscles were so sore I couldn't straighten up. I looked like a hundred-year-old hunchback. I spent most of the next two days walking around looking at the floor instead of where I was going. Embarrassing! The point to all this: I discovered a whole lot of muscles I never dreamed I had.

Depending on which expert we might consult with to count our muscles, our numbers could range from 656 to 850. That's because some muscles are distinct and others are part of more complex muscular structures.

It goes without saying that muscles are extremely important; in fact, without them motion of any kind would be completely impossible. Your muscles turn body energy into motion for you. Communication and expression are only possible when your muscles make it happen, whether it's speaking, seeing, making hand gestures or moving your body.

What exactly are muscles? They are bundles of long cylindrical-like cells called fibers that act in groups and can only contract to tighten or relax to loosen. They range in size from 1 to 40 microns (that's millionths of a meter) long and from 10 to 100 microns across. Their contractions are triggered by electrical stimuli, whether from within the body, as from a nerve cell, or externally from some kind of electrical shock or stimulus.

A close study of the body reveals that muscles can be categorized into four distinct shapes:

- Flat Muscles, like those found in your forehead or your diaphragm.
- Circular Muscles, also known as sphincter muscles, like those at the entrance and exit of your stomach.
- Spindle Muscles, which look like spindles—thick in the middle, but tapering down at each end. Biceps and triceps are good examples.
- Triangular Muscles, thick muscles that are major workhorses in your body. A good example would be your deltoid muscles, the thick shoulder muscle that moves your arm out and away from your body's midline.

We'll look at the larger topic of muscles the same way a biologist or kinesiologist would—by grouping them into their three distinct categories: cardiac, skeletal and smooth. And not to be left out, there are facial muscles and your tongue; we'll look at them briefly, too.

CARDIAC

Heart muscles are called cardiac muscles, and they exist nowhere else in your body than your heart. They set their own rhythm, and you never need to give them a second thought. Their function is simply to contract and release, thereby driving the pumping action of the four chambers in your heart as it circulates your blood throughout your body. You can refer to the earlier section on the circulatory system for more information.

Think of the task the heart has to perform: it truly is the hardest working muscle in your body as well as the one that never rests. During your lifetime your heart will beat 2.4 billion times. Its pumping action is forceful enough to squirt blood close to 30 feet (9 meters). It wins the prize for endurance hands down.

Here's a true anecdote to add a little humor to this last claim. When I was in university our faculty all headed off one day to the local Red Cross blood donor clinic. One wise guy (no, not me) took an IV bag and filled it with chicken blood, then stuck it under his shirt with a plastic tube that he taped to the opposite arm from the one he would give blood from. After the real blood hookup was running and the nurse had left, he tilted his other arm so the tube was pointing up, gave a loud shriek, then squeezed his arm to his side, compressing the IV bag, sending the chicken blood squirting meters into the air. Needless to say, the nurses went into immediate panic mode and the rest of us were in hysterics. The truth was discovered. The dean was contacted. The sky fell in!

SKELETAL

Simply put, skeletal muscles attach to your skeleton and are your body's engine that makes motion possible. Unlike the involuntary action of your cardiac muscles, your skeletal muscles need you to tell them when to contract and release. That ball isn't going anywhere unless your brain knows you want it kicked and passes the appropriate instructions to your muscles.

Skeletal muscles come in pairs. Hold your left arm out horizontally, palm up. Curl your arm up and touch your shoulder with your hand. Move it back horizontally. You've just used one set of muscles, your biceps, to move your arm up and another to move it down. Put a weight in your hand and repeat the above. The difference between the two is the signal sent from your brain down a nerve path to your muscle, telling it how much to contract in order to achieve the required motion.

SMOOTH

Smooth muscles contract involuntarily, without your input, thank you very much. Unlike skeletal muscles that contract along an axis, shortening in a smooth muscle takes place in all directions at the same time and with a greater staying power than skeletal muscles have.

They can be found an all kinds of places within your body. When it's time to pee, that's smooth muscles in action; when they contract, they squeeze your bladder to push the urine out. Most internal organs, like your stomach and intestines, are lined with smooth muscles. They are also the agents within a woman's uterus that contract to complete the birthing process. When your eyes change focus, smooth muscles are at work. They also help move chyme—food leaving the stomach—along as it passes through your digestive system.

FACIAL MUSCLES

You've probably heard it said that it takes a lot more muscles to create a frown than it does a smile. Most of us are not nearly as concerned about our facial muscles as we are about frown lines, yet our faces have many muscles. Be honest. Have you ever stood in front of a mirror and made goofy faces? It's quite intriguing how many contortions our facial muscles can support. Although some facial muscles attach to the bone like skeletal muscles do, many attach under the skin. This gives rise to an endless variety of faces you can make in the mirror. I ask you, how would we communicate properly without facial muscles?

YOUR TONGUE

Did you know that your tongue is actually a group of muscles working together but only attached at one end? Talking and chewing food are the most common activities in which it is involved, but it also has one other dubious characteristic: it is the most unruly muscle of your body. Want proof? Consider how it can spontaneously and unexpectedly spew forth expletives when something or someone irritates you! Why, it can ruin a friendship in a heartbeat.

Here are some questions. Without muscles in the first place, life would never have had a beginning. Were they a part of the original life form, and, if not, how did life endure until evolution required them? And when did the connection between muscles and the brain develop, up front or later on?

Here's some muscle trivia to end this section:

- The muscles you are born with are all you will ever have, so care for them.
- Really muscular people have the same number of muscles you do, just a lot thicker.
- Your tongue is your strongest muscle.
- There are more than twenty different muscles in your hand alone.
- Your leg muscles will carry the average person some 5 million steps a year.

TENDONS

Have you ever had two completely dissimilar things you needed to join together, like wood and glass? Trying to find the right adhesive can be a challenge. Way back in the primordial swamps in the land-before-time, there was no magic bonding agent. This created a challenge in what to use to join muscles to bones. Enter the tendon. However it worked out, muscles connect to bones via tendons. Without tendons, you would be a rather pathetic blob, unable to move at all like you do today.

Think about what gelatin looks like, only harder and not as elastic. That's what tendons are like. Within your body there is a type of protein called a collagen. If you could see millions of these fibrous collagen proteins under a microscope you would see how they weave together to form a strong strand of flexible tissue. This is a tendon that literally grows into the muscle at one end and into the bone at the other, making a connection that is quite difficult to break. But possible it is! And I proved it one day much to my dismay.

I learned the hard way some years ago how very painful it is to tear a tendon or to rip it loose from the bone. One fine day, I decided to enter a swimming pool and time myself in a speed trial, with no forethought to building up my elbow muscles in advance. I just wanted to see if I could break my old record that I had set twenty-five years earlier, and I only had a short amount of pool time to do it in. My resulting lap time wasn't too bad, but the elbow pain left me almost immobile for weeks. Dumb move! I tore the tendons off the bones in each elbow and even to this day they haven't properly reconnected.

LIGAMENTS

This leaves us with one more component to look at: ligaments, a kind of bodily Crazy Glue, if you will. Somewhat similar to tendons, ligaments are also fibrous connective tissue, but they connect bone to bone, not muscle to bone. They support bone joints and prevent motion in the wrong direction.

Have you ever popped a joint, like in a dislocated shoulder? Ouch! If you did, you likely stretched the ligaments that keep your arm bone in your shoulder socket beyond their limit.

Do you do stretching exercises? If so, you are stretching your ligaments, and that makes you more limber. But for all that stretching activities achieve, your ligaments only extend to a maximum of about 4 percent.

So what about tendons and ligaments and their ability to connect dissimilar structures: when did they come onto the scene? Were they forethoughts or afterthoughts? And if afterthoughts, how did humanity survive those interim millennia without them?

What Goes In Must Come Out

The Digestive System

Someone once said, "Just because I don't understand how the stomach works doesn't mean I can't eat." We could all say that we basically know what happens when we eat something, but let's go just a slice deeper and explore the processes that we are either unaware of or have taken for granted, and see what additional insights we can mine about this fantastic machine we call the body.

Within the digestive tract there are digestive secretions and six principal organs—the tongue, pharynx, epiglottis, esophagus, stomach, and intestines—and some other essentials, like lips, teeth, sphincter muscles, and even our sense of smell. We could also talk about digestion really starting with the eyes or the nose, as the perception of food gets initiated and the saliva builds up in readiness for digestion. We'll talk about that shortly, but for now, let's begin at the mouth, with its various key parts.

THE LIPS

It has been said that the lips are often the first feature we notice about a person. But look beyond the appearance and we make some interesting discoveries.

Lips not only have their own muscle groups but also share muscle groups with the mouth and cheek. Of all the organs in the human body, the lips are capable of one of the highest degrees of mobility. Consider just a few of their motions: manipulating food as it enters the mouth, holding food and fluids in the mouth, forming facial expressions, articulating sound and speech, and, of course, kissing (my personal favorite; my wife's too). Were it not for lips, suckling for the newborn would render breastfeeding impossible, and if this had been the case from the beginning of humanity, the ramifications would have been disastrous, wouldn't you agree?

TEETH

Fire up your imagination. You're in a steak house and the filet mignon arrives at your table. It's perfect! Deliciously, scrumptiously, mouthwateringly perfect! You grab your steak knife, slice a hunk of meat and pop it into your mouth. You chew, you enjoy, and you savor the moment. Aaaah! What could be better?

New image. You're visiting friends for dinner and they serve a platter of succulent, spicy chicken wings. No steak knife this time. What do you do? What else? You grab a wing, bite off a piece and continue to chew, to enjoy, and to savor, don't you? Sure! So what?

Well the "so what" is really a "so how." You see, with the chicken, your front teeth were the equivalent of the steak knife—sharp, chisel-shaped to cut through the meat easily.

Here's what we so often take for granted: the front teeth, twelve of them, are used to cut and tear our food, while the rear teeth are used to grind it. What if this had not been the case in the earliest man? What if their front teeth had not been sharp? What if they had all been grinders? It would have been like trying to slice through that steak with a blunt blade a quarter of an inch or more wide! Come to think of it, have you ever seen a replica of earliest man that didn't have sharp front teeth? I haven't.

TONGUE

OK, the food is in the mouth; what's next? My parents told me that my first word was "No," quickly followed by my first complex phrase, "I won't," and then my more deterministic pronunciamento, "I'm not gonna!" Such communication would not have been possible were it not for a multi-talented little muscle we call the tongue. It achieves all kinds of essential things:

- With it we move food back and forth in our mouths, mix it with saliva, massage it into just the right-sized little ball and then position it to the back of the throat for swallowing. This is the essential first step in digestion.
- The tongue, along with the throat and the roof of the mouth, contains about 9,000 taste buds, without which everything would taste the same. Restaurants would be out of business in no time at all!
- We use it to form sounds, without which spoken communication would be impossible.
- We can also use it to communicate visually, be it rudely (as in sticking your tongue out) or otherwise (as in licking your lips after a great meal).

One more point: what we put in our mouths, whether food, fingers, pencils, liquids, etc., can often come with bacteria and other unsavory (no pun intended) stuff hitchhiking along, which can infect the tongue and possibly damage its surface. Yet the body has its own special solution to this end—a mucous membrane that completely covers and moisturizes the tongue. How ingenious! When do you suppose that appeared on the scene?

SALIVA

As food is chewed it is mixed with saliva (2 percent electrolytes, mucous, antibacterial compounds and enzymes, 98 percent water). We each have between 600 and 1,000 minor salivary glands; considering the size of the mouth, that's a high concentration of glands. Saliva serves many purposes:

- Lubricates food: As you chew food into smaller and smaller pieces, saliva lubricates and binds the food into slippery small round masses that will slide down your esophagus without harming or directly touching its mucous walls.
- Solubilizes dry food: If your food didn't exist in a soluble state, you couldn't taste it.
- Aids in oral hygiene: Your mouth usually feels and tastes terrible when you wake up. That's because your flow of saliva is much less at night than during the daytime. While awake, your mouth is constantly being flushed and lubricated with saliva, persistently removing leftover food debris and generally cleaning the mouth.
- Initiates starch digestion: Enzymes in your saliva begin to digest dietary starch before it passes to your stomach.

- Maintains teeth: Saliva is an important source of calcium and phosphate ions essential for normal tooth preservation.
- Enhances taste: Without saliva to act as a medium for holding and transporting dissolved and suspended food materials that chemically stimulate taste buds, you would have a greatly reduced sense of taste.

Without salivary glands, you would quickly become very ill and unable to disinfect your mouth, moisten food for digestion or convert glucose. Your teeth would rot, and your food intake would rapidly diminish to nothing. In short, you wouldn't survive for long. Without saliva from the very beginning of human life, humanity would have been short-lived. Wouldn't this suggest that this functionality was available from the very beginning?

PHARYNX

Once food passes out of your mouth, its next stop is your pharynx, located directly behind your mouth. Your pharynx plays a dual role in the body: it is part of the digestive system as well as the respiratory system. It links your mouth to your esophagus, a narrow muscular tube about 10 inches (25 centimeters) long that ends at your stomach's top orifice. Running in parallel with the esophagus is your windpipe, which ends in the lungs.

Have you ever considered this? If you pinch your nose tightly, you can easily breathe in and out through your mouth, right? But you also swallow through your mouth. So how does all that work? Why doesn't our food end up in our lungs? The answer is just beyond the pharynx, and it is called the epiglottis.

EPIGLOTTIS

Because both air and food pass through your pharynx, there has to be some way of preventing food and liquids from entering your windpipe (trachea) and lungs. If that were not the case, you would quickly choke to death. Chances are that at one time or another you've had something you ate head for your lungs and not your stomach. When it does, it can be a horrifying experience, sometimes even needing a severe rescuing action like the administering of the Heimlich maneuver to dislodge the food back into your mouth for expulsion.

The solution is a flap of connective tissue called the epiglottis. It is located at the back of your pharynx and closes over your windpipe when food is swallowed, thereby preventing choking. Even more amazing is that you never need to think about the epiglottis; it is completely controlled by your autonomic nervous system, which automatically sorts out whether air or food is passing. That is quite a remarkable feat! If this were an evolutionary thing, how did the early generations survive without the ability to direct air and fluids in the right directions?

ESOPHAGUS

The next station for food and fluids is your esophagus, a 10 inch (25 centimeter) tube leading to your stomach. That short journey will take about seven seconds. But wait. Why wouldn't it be almost immediate? After all, a ten-inch vertical drop isn't very far. The answer is that your esophagus has two layers of muscles, and they produce a wave-like motion, called peristalsis, which moves the food slowly but efficiently along its way to your stomach. This is why you can recline while eating a snack and your food will move along your now inclined esophagus. There will be more on peristalsis later.

STOMACH

Up to this point in the digestive process, saliva has prepared your food for two main events: primary digestion in the stomach and final digestion in the intestines.

As food enters your stomach it is further broken down and thoroughly mixed with gastric acid, pepsin and other digestive enzymes, which break down, or denature, proteins. Enzymes in your stomach work best at a specific pH level[6] and temperature. The primary function of your stomach acid is to set an optimum pH level for the reaction of the pepsin enzyme; it has no role in breaking down food, but it does kill many microorganisms attached to the food. As digestion in your stomach continues, various vitamins and other smaller molecules pass through your stomach membrane and are directly absorbed into your circulatory system.

INTESTINES

Within your stomach, food is in a semi-liquid state. Upon exit into your small intestine it is known as chyme and consists of partially digested food, water, hydrochloric acid and various digestive enzymes. The majority of digestion and absorption occurs here. This path, from the stomach to the colon, is about 20 feet (6 meters) long and about 1 inch (2.5 to 3 centimeters) in diameter. Its internal lining is anything but smooth; in fact, its complex structure if unfolded and laid flat would have a surface area of approximately that of a tennis court. Without this vast surface area, your small intestine would not be able to assimilate all the essential nutrients passing through it and your health would quickly waste away.

Various nutrients are further broken down chemically into smaller molecules to allow absorption into your circulatory or lymphatic systems. Peristaltic waves transport this processed food throughout your entire intestinal tract until the waste portion—that which is not absorbed into the intestinal wall—is ready to exit your body.

Without peristaltic waves to move food through the digestive system there would be no intestinal absorption of nutrients, and the intestine would immediately plug up and the individual would die. Once again we need to consider the question that just won't go away: was this capability there from the beginning?

KIDNEYS

So far we've been focused on the digestion process as it relates to the solids we eat. It's time to switch to the liquids. The kidneys are the vehicles by which soluble wastes get excreted from the bloodstream in the form of urine. Urine contains a range of substances that vary with what is introduced into the body, including excess water, sugars, urea, toxins, inorganic salts, organic compounds, proteins, hormones and a variety of metabolites. In this respect, they perform a waste removal function.

They also have several important secondary functions, including the regulation of electrolytes, the correct balance of acids and bases, and the production of hormones, renin (an enzyme that regulates the body's arterial blood pressure) and erythropoietin (a hormone that controls red blood cell production).

We know life can be sustained on just one kidney. But what if there is disease of both kidneys? If you know anyone unfortunate enough to require renal dialysis, you'll know that it is a life-or-death situation. And what if the body had every organ it required from the very beginning except kidneys? Then how long would life have existed without the ability to purge excess fluids and hazardous leftovers?

BLADDER

I'm constantly fascinated by the architecture of the body. I read in a medical journal that the bladder "sits on the pelvic floor." In the sketch that accompanied that comment I could see clearly that the bladder was considerably lower than the kidneys. So what? So, gravity is a wonderful thing, isn't it, and having the bladder lower than the kidneys takes great advantage of that. What if the bladder were higher than the kidneys?

On the other hand, if there were kidneys but no bladder, all fluids would be on a direct pass-through, and that would make life a lot more complex and restricting, not to mention just a wee bit embarrassing (sorry, I couldn't resist the pun). If you think long road trips are a license for frequent gas station stops, what would life be like without a bladder? I dare say that meetings would be shorter and washrooms would be larger! Enough said.

SPHINCTER MUSCLES

Before departing this section, I must introduce this indispensable muscle type we find in at least forty different locations throughout the body. A sphincter is a ring of muscle, usually circular in shape, surrounding and serving to guard or close an opening or a tube. It relaxes and opens as required by normal physiological function. In the context of the digestive system, there are several sphincter muscles.

THE UPPER AND LOWER ESOPHAGEAL SPHINCTERS

Remember, the esophagus is the tube that carries food from your throat to your stomach, and these sphincter muscles act as valves, one just below the junction of your throat and esophagus and the other just before the junction of your esophagus and stomach. When there is no food passing through your esophagus these muscles contract, preventing a mixture of digesting food and stomach acid from flowing back up to your mouth and seriously damaging the lining of your esophagus in the process. When there is food to swallow, these muscles relax, allowing food to reach your stomach. Think about it: how is it that our bodies have all these awesome and essential control features that happen completely without our intervention? What are the odds?

THE PYLORIC SPHINCTER

Just as the lower sphincter allows food to enter your stomach but not flow back into your esophagus, the pyloric sphincter has a similar function at the other end of your stomach, acting as a valve to prevent chyme—already processed in your stomach and now passed into your small intestine—from being regurgitated back into your stomach.

THE URETHRAL SPHINCTERS

I think you'll soon agree that these have to be among the most essential muscle groups in your body, because without them life would be quite messy and embarrassing. These muscles surround the urethra—the tube that leads from your bladder to the outside of your body—and constrict or release to control the flow of urine from your bladder. Think what life would be like without them; OK, that's long enough; I'm sure you have a very clear picture. What do you think; were these muscles a part of the original gooey package?

THE ANAL SPHINCTERS

And last, but certainly not least, there are two sphincters that occupy the 1 inch (2.5 centimeters) or so of the anal canal and tighten or ease to control the release of feces from the body. I'm sure you would agree that they, too, fall into that group of muscles deemed to be among the body's most essential. Again, would you consider these to have been an integral part of the original package? In fact, what would the ramifications have been if there had been no exit at all for feces from the body? Don't spend too long thinking about that one; just ask yourself if this, too, was part of the original design or a later add-on.

PERISTALSIS

Have you ever been to a stadium or arena to watch a major event, one with lots and lots of fans? There are people along the routes into the event that direct your progress. "Keep moving. Move right along," they say. You know, your body has a very similar function that takes place throughout your digestive tract: peristalsis.

Everything you do, up to the point you pass the food into the back of your mouth and swallow it, is voluntary; you are aware of what you are doing and initiate the actual swallow. From then on, the process becomes involuntary; the nerves in the esophagus that control the process are stimulated by the food you have swallowed resulting in a series of smooth wave-like movements of the walls of your esophagus nudging your food down to your stomach. This same peristaltic motion can be found elsewhere; it helps move urine through the tubes that connect your kidneys to your bladder, bile from your gallbladder into your duodenum (upper intestine), and chyme throughout your intestine.

Isn't it wonderful how our digestive system not only processes our food intake to get the nourishment we need from it but also has this "move along" capability to prevent food from jamming up within us, putrefying and leading to disease and death? Equally interesting is the fact that these peristaltic waves always move in the right direction, never in reverse, and never randomly this way and then that. If this weren't part of the original package, how long did humankind have to struggle to survive until it evolved? Could these be more signs of intelligent design? I wonder.

Good Guys vs. Bad Guys

The Immune System

Allow me to introduce this section by way of a short allegory. A so-called sage decided to design the most ingenious city ever dreamed of. He meticulously planned its myriad parts to fit perfectly and harmoniously with each other. Although its population would be small to begin with, its design could effortlessly expand to accommodate much greater numbers. Everything the city would need to survive would be both available and accessible to it, supported by ingenious delivery and handling systems. Nothing was redundant. Most impressive of all, as the city aged and its infrastructure needed to be replaced, all that would happen without intervention by its occupants.

When the design was complete, the sage set about to construct his Utopia. And so it came to be. In true utopian fashion, this perfect city had no natural defenses; in its perfect world, why should it? One day, a stranger entered the city and noticed how wonderful a place had been made for him to live; everything he needed was available to him. But soon after his arrival, many things gradually ceased to operate as they had been designed to. The citizens lost their easy access to food and water; the road systems became clogged with unexpected traffic and the air changed both in color and smell. In their vernacular, something definitely "un-good" was happening.

Once the city elders realized that the only possible reason for this change lay in the arrival of the stranger, they started to question him, asking him where he came from and what his name was. To this the stranger replied, "I come from a place called Outside, and my name is Decay." You see, our sage had never considered the necessity of having any defenses to protect his jewel, and the first outsider to venture in initiated its demise.

Our bodies are like that city, but with one significant difference: we do have a defense system. We call it our immune system. For the most part, it does a stellar job, far beyond the understanding of all but a few of us. Like every other system we have looked at, this topic has complexities and depths of detail far too deep for this type of book to even consider tackling. So let's see what we can learn from a thin slice of understanding just what the immune system is and how it works.

If you were to ask your doctor for a terse definition of the immune system, you would likely get a reply that categorizes it as a collection of biological processes that detect and kill pathogens, those dangerous substances or cells that invade the body and can potentially cause disease. Functionally, your immune system removes wastes, cellular toxins, dead blood cells, and cancerous cells and, in conjunction with your circulatory system, delivers three essential substances to your cells: hormones, oxygen and nutrients.

When diseases and other microorganisms (germs, viruses, cancer cells, parasites, bacteria, etc.) that invade your body are considered, the role of the immune system becomes increasingly complex. Not only must its agents be at the right place and at the right time to attack the invaders, but they must also be able to differentiate them from your body's own healthy cells and tissues. Any cells that your body identifies as invaders and potentially dangerous will stimulate your immune system to respond to defend those healthy cells by attacking and eradicating the invaders. As if this is not tough enough, pathogens can evolve, creating a whole new kind of threat, perhaps not yet recognized by your immune system's defenders. Your immune system possesses the innate ability to adapt to counter these new pathogens.

This raises a question worth considering. Was this adaptive nature there with all its wondrous capabilities from the very beginning? If not, then the first killer pathogen to come along would have wiped out mankind in quick order.

Putting an End to Waste

The Lymphatic System

Your body has two major fluid systems intricately linked and intertwined within your body that serve to transport your immune system's agents: the blood and the lymph. Consider the latter.

LYMPHATIC SYSTEM

That's a term I had heard for years and used it, on many occasions I'm sure, without really knowing what it was. When my wife was diagnosed with breast cancer and required a lumpectomy, her physician also removed a number of lymph nodes. That's when I began to understand what this collection of organs—lymph nodes—is all about and how absolutely essential it is to the life of the body.

Every cell and tissue in your body needs continuous feeding with nutrients and continuous irrigation and drainage of waste products. Your lymphatic system is an integrated network of some 500 to 600 bean-shaped lymph nodes—lymphatic fluid filters, if you will—situated throughout your body. Blood vessels feed the tissues that bring the nutrients to your cells, while your lymphatic system carries away the wastes. Without this essential component of your body's drainage system you would not be able to filter out harmful toxins and bacteria from your cells, leaving your body highly susceptible to disease. The greatest densities of these nodes are found in your underarms, chest, neck and abdomen. Unlike your blood system, which is circulated by your heart, your lymphatic system has no pump and must rely on other means to enable its flow. As your skeletal muscles contract, they cause your lymph fluid to move through its vein- and capillary-like channels. Because there is no back pressure in these channels, your lymph vessels, like your veins, are remarkably endowed with one-way valves that prevent backflow.

Each node contains lymph, a colorless fluid with a high white blood cell concentration. Lymph drains wastes from the body's tissues, routing them through the lymphatic system's network of channels and into the blood system. I watched the effect a reduced number of lymph nodes had on my wife's health after her chemo and radiation therapies; I watched and I grieved. The part of her body that now had a marginalized lymph drainage system took much longer to heal than if the lymph nodes had not been removed, not to mention having other significant collateral problems.

When we look at our bodies, it is abundantly clear that there are systems for delivering nutrients and there are systems for removing wastes. Would you agree that without both of these systems in place and functioning from the very beginning, life would have been impossible?

Your Own Personal Airlines

The Respiratory System

"I can't breathe! I need air! I'm gonna die!" Have you ever felt this way? I have. Some years ago I was at a French Club Med on the Caribbean island of Martinique and often snorkeled with the club's scuba diving instructor, Jean-Paul. One day he challenged me to "dive to the bottom" with him. Macho-me was up to the challenge. Problem was, the bottom was in the shadow of the boat, and I couldn't tell how deep it was. I asked him, and he said "Vingt-neuf." "Twenty-nine feet," I thought; "Oh, I can do that!" I'd done free dives to that depth before.

So off we went, Jean-Paul first, then I a fraction of a minute later. Ten to fifteen seconds into the dive I noticed that I was actually starting to sink; my buoyancy had become negative as the depth increased. That was good; it speeded up the descent and shortened the time to get to the bottom. But wait, the bottom was still a long way below me. It didn't take long to figure out that this was no shallow dive. It wasn't twenty-nine feet at all; it was twenty-nine meters! Ninety-five feet! Gasp!

I had a long way to go. Seconds before I got to the bottom, I noticed Jean-Paul grabbing a stone and stuffing it into his bathing suit. Good idea, I thought: evidence! I better do the same. Euphoric, I finally hit bottom, grabbed a stone and poised myself vertically to start back to the surface. Keenly aware of the depth and my aching lungs, I fought off fear and started to swim slowly and forcefully for the surface.

All of a sudden the clarity of the water became a blur. What was happening? I looked down and was still on the bottom, my flippers stirring up the sand. I wasn't ascending! My strokes weren't powerful enough to overcome my negative buoyancy. Oh help! Instinctively, I sunk back to the bottom and then pushed off as hard as I possibly could, trying to find the right balance between raw fear, my burning lungs, and the temptation to exhale some of what little air I still had.

Getting back to the surface seemed to take forever. It's true what they say about your life flashing in front of you when you are suddenly in a life-or-death situation; it was starting to happen to me, and I realized how foolish a situation I had put myself in.

By the time I broke the surface, my lungs felt like they would explode. I had blood in my nose, eyes and ears. I'd been under almost two minutes. That first gulp of sea air was the best air I've ever tasted. If you're saying, "I bet he never did that again," you got that right. And if you said, "Dumb," you got that right, too.

The point being? The point being, most of us are healthy enough that we take our air intake very much for granted. Think about how elegantly simple yet astoundingly complex our respiratory process is.

We each have what is commonly called a "diaphragm." It's a muscular sheet that extends across the bottom of our rib cage. When our brain recognizes the need to inhale, it sends a signal to our diaphragm, instructing it to contract. This increases the size of our thoracic cavity (chest, if you prefer), reducing its internal pressure, thereby creating suction. This causes our lungs to expand, resulting in air being drawn into them. When it's time to exhale, our brain sends a signal to our diaphragm to relax. Our chest contracts and our lungs return to a normal volume, causing the air within to be expelled.

If we consider the simple mechanics of air in, air out, this is rudimentary physics. But we're not just dealing with a machine here; we're dealing with a physiological system that is totally automatic, working whether we give any thought to it or not. Someone once said that the two most powerful forces in nature were the instinctive drives to breathe and to procreate. Most of us spend next to no time thinking about one of them and a lot of time thinking about the other. Enough said.

During the inhalation and exhalation process, complex events are taking place at two different levels: breathing and cellular respiration.

BREATHING

Breathing is the physiological process of exchanging carbon dioxide—the product of cellular respiration—for a fresh supply of oxygen that your circulatory system then transports throughout your body. This is the complex part I mentioned a few paragraphs back. To distill this to its essence, here's what happens. Something has to make the air exchange happen, and that's the job of tiny air sacs within your lungs called alveoli. Your two adult lungs have 300 to 500 million of them, each surrounded by a web of capillaries—microscopic blood vessels no more than a single cell thick. If you could add up the surface area of all these alveoli, it would be about the size of a single tennis court. That's about eighty times the surface area of your body.

Your circulatory system capillaries supply blood to each of your alveoli. Here, blood wastes and carbon dioxide are exchanged for a fresh supply of oxygen, which, in turn, is returned to your circulatory system to complete the cycle of replenishing your cells with oxygen.

CELLULAR RESPIRATION

Cellular respiration is a series of dual processes that first release biochemical energy within your cells for metabolism and then remove waste products from them.

Most adults breathe twelve to fifteen times a minute—roughly 17,000 to 22,000 times a day—but we never need to think about it, because it is all under the control of our autonomic nervous system. Even if you try holding your breath for too long, your body will detect the carbon dioxide buildup, forcing you to exhale and to start to breathe again. That's what was happening to me as I hit the surface having come up from my deep dive.

Here are some other factors you never need to think about that can influence your breathing rate:

- If your blood's oxygen concentration gets too low, your depth and rate of breathing will automatically increase.
- If your blood's carbon dioxide level gets too high, the respiratory centers will automatically trigger

deeper and faster breathing until an acceptable ratio of carbon dioxide to oxygen has been restored.

- If your blood becomes too acidic, your respiration will increase.
- Just as not over-inflating a balloon can prevent it from rupturing, so too your respiratory centers will detect lung over-inflation and trigger exhalation.
- During periods of high stress, pain or emotion, your brain will signal your respiratory centers to adjust your breathing rate accordingly.
- Sneezing is your respiratory center's method to quickly eject irritants such as smoke, dust, pollen, or unpleasant fumes from your lungs.

Producing regenerating, sustainable life from the get-go is not a question of "try it and if it doesn't work we'll modify it and try it again." If it's not right the first time, there likely won't be a second time. If all aspects of our breathing systems didn't integrate and behave perfectly from the very beginning, what do you think the chances of life would be?

Some of Life's Necessities

The Glandular and Hormonal Systems

I remember getting the mumps when I was a kid; my throat swelled up like I had an orange stuck in there. My parents tried to allay my fear by brushing it off as a glandular problem. And it worked. I had no clue what a gland was, but the doctor said the one that was causing my mumps was no big deal and that, in fact, I had all kinds of glands in my body. I felt reassured.

As I began to contemplate this section, I started to consider just what "all kinds of glands" really means and whether they are important or not. The answer was far beyond what I had expected. First of all, what is a gland?

A gland is an organ that produces a chemical secretion that gets released either into the bloodstream, into the body's inner cavities, or onto its outer surface. There are dozens of glands in your body, each playing a key role in your continuing well-being. Depending on your gender, you will find them in your areolae, breasts, cervix, digestive tract, duodenum, esophagus, eyes, eyelids, heart, hypothalamus, intestines, kidneys, liver, lymph nodes, mouth, nose, ovaries, pancreas, parathyroid, penis, pineal, pituitary, placenta, respiratory tract, skin, stomach, striated muscles, testes, thyroid, tongue, urethra, uterus, vagina, vocal chords, vulva, and many more.

Within this family of glands there are specific glands that are categorized as making up the hormonal system, also known as the endocrine system; their sole purpose is to produce hormones. I realize there are a lot of stories about hormones and the sexes, so let me add a little bit of background to make these stories all the more meaningful. The word *hormone* is derived from the old Greek word *hormao,* which means "I stir up" or "I set in motion"; either way, it conveys the sense of getting your vital juices flowing.

Hormones are substances created by organs in one part of your body that send chemically coded messages to other parts of your body via your bloodstream. They instruct those receptor organs to behave in a particular manner. Although your body produces a variety of hormones, each one communicates only with the cells that have been genetically coded to react to it. The specific hormonal glands are the adrenals, amygdala, hypothalamus, pancreas, pineal, pituitary, ovaries, testes, thyroid and parathyroid. Individually, each serves an extraordinary function. Here are just a few.

ADRENALS

When I was in university, I had a motorcycle. One day, as I was traveling up a major roadway (fast—as in speeding), I noticed a car parked well off on the shoulder of the lane in which I was riding. There was no hint of any motion or intent that its driver would enter my pathway, so I maintained my speed.

Suddenly the car roared into life and turned left directly in front of me, leaving me absolutely no way to avoid crashing into its side. And that's exactly what happened; however, in that instant when I knew that impact was inevitable and only a fraction of a second away, my adrenal glands must have kicked into overdrive. Instinctively, and with an unnatural strength, I jumped straight up and off the pegs my feet rested on, and I catapulted over the roof of the car without even touching it. Even better, I landed in soft grass on the shoulder of the road, slid about fifty feet and was totally unhurt, just hugely ticked off at the driver of the car. That adrenaline rush saved my life.

You have two adrenal glands, one on either side of your body, sitting directly on top of each of your kidneys; that's why they are called adrenal (*ad,* "near" or "at," plus *renes,* "kidneys"). They produce hormones that help us deal with the variety of daily stresses we all endure, like motorcycle crashes, getting married, getting unmarried, taking a holiday, being so tied to our jobs that we don't take time for a holiday, going on a business trip, writing an examination, having an unexpected hospital stay, coping with traffic, going shopping with our spouses, etc. Would life's earliest generations have been self-sustaining without this powerful chemical racing through their bloodstreams like greased lightning at just the instant it was needed?

AMYGDALA

First of all, it's pronounced *am-ig-dah-lah,* and it's part of your body's limbic system, the system that hosts and controls sexual arousals, motivations, fight-or-flight reactions, and emotions, to name just a few. The most common emotions associated with the amygdala are those of anxiety, anger, fear and avoidance.

Do you remember your early school days when you had to give a speech to your classmates? Your palms would sweat, your legs felt like rubber, your heart pounded so hard you thought it would explode, your antiperspirant had abandoned you, and you just knew every eye was riveted on you, waiting to see how long it would take you to mess up big time. You were in a state of fear, anticipating the worst. Your amygdala was at work, decoding all the emotions you were feeling and preparing you for a response.

If you didn't have such a capability, you'd never be able to sense danger or to respond rapidly to threatening situations. If early humans didn't have this built-in "intuition," wouldn't they have been in mortal danger the first time a hungry animal came along? Such an attack could have terminated civilization.

GALLBLADDER AND BILE

The word *gallbladder* does not have the most appealing sound, but without this gland you couldn't digest the food you eat. Here's how it works: your gallbladder is a reservoir for a substance the liver produces called bile, which it stores to have ready for secretion when you eat a meal. Bile breaks down fats so your body can metabolize them.

When you ingest food, especially fats, this triggers your gallbladder to relax its associated sphincter muscle so bile can enter your small intestine. At the same time, your gallbladder begins to contract, causing its pent-up bile to squirt into your small intestine, where it helps to emulsify (i.e., break down) the fats in the foods you've just eaten. Without bile, your ability to digest food would be far from beneficial, and the toxins in your body would quickly reach unhealthy levels. This is not a path one would want one's body going down, wouldn't you agree?

HYPOTHALAMUS

The hypothalamus is a region of your brain—about the size of an almond—that produces hormones to regulate bodily processes such as body temperature, hunger, thirst, fatigue, sex drive, sleep, the release of hormones from other glands, and circadian cycles.[7]

Its main function is to maintain your body's status quo, which is often medically referred to as the "set point." We experience this set point all the time but are usually oblivious to it. Factors such as fluid and electrolyte levels, body temperature, and blood pressure are maintained from day to day within superbly fixed parameters. Our hypothalamus receives constant input on our body's status from a number of sources and then initiates whatever changes are necessary to bring that status back to within some reasonable tolerance of its set point. If this kind of automatic control were missing, our bodily systems could quickly run amok, perhaps fatally. What do you think the probabilities are that this incredible little chemical computer was part of the original human package?

MAMMARIES

I recognize that most of us will have an acceptable understanding about the female breasts and their role in producing milk for newborn babies. But here's something I didn't know until I started this research. At birth, the first fluid a newborn receives from its mother's breasts is a liquid called "colostrum." It's the infant's initial source of nourishment until the mother's milk starts to flow. What is so awesome about colostrum is that its more than ninety known components provide vital support and nutrients to the newborn in areas such as improving liver function, boosting the immune system, aiding in proper digestion, decreasing allergy symptoms, helping regulate moods, aiding regulation of normal blood glucose levels, assisting in the regeneration of the muscles and bones, increasing concentration, and acting as a mild laxative to help the newborn's immature digestive system pass its first stool.

One of the references I researched on this subject gave all the credit to "Mother Nature." Good ol' Ma Nature, whoever she is, really must have had it all figured out before the first signs of life. (For the more inquiring minds among my readers, did you ever wonder who Father Nature was and what role he had in all this? More on this later.)

We live in an unbelievably complex structure we call our body. How would we ever have gotten out of the gate without nourishments like colostrum?

OVARIES

When my daughter was born it was still the exception for the father to be present during the delivery. But there I was. For me, the miracle of birth is awe-inspiring. Now, several years later, my daughter has three daughters of her own and my son has one. During the late second and early third trimesters, these little sweeties each had somewhere near six million eggs each in their ovaries, a number that diminished to a million or so by the time they made their grand entrance into the world. Think of it: my daughter was not only carrying her children, she was carrying her grandchildren as well. By the time each child reached puberty that number had further dwindled to a quarter of a million or so, but still, that's more than enough eggs to last for her fertile years. To put a twist on an old saying, "You can count the number of eggs in a woman, but you can't count the number of women in an egg."

A woman will ovulate around four hundred times during her reproductive years, with one egg normally being released during each of her menstrual cycles. Each released egg will be one of the same eggs she had as a fetus while in her mother's womb; no eggs are created after birth. That makes the egg one of the longest surviving cells in the mother's body.

So now we face the inevitable questions. Is it plausible to think that the big ol' lightning bolt created an immediate capability to produce offspring from that first "creation"? And when you think of a woman's monthly cycle and the ovulation that occurs, and the number of times she may have sexual intercourse before there is a mating of sperm and egg, what are the chances that her ancient ancestors were endued with enough eggs to last them through their entire fertility period?

PANCREAS

This leaf-shaped gland is about 6 inches (15 centimeters) long, lies beneath your stomach and is connected to your small intestine at the point where it connects to your stomach. Part of your digestive system, it has two main functions: one to produce and secrete enzymes essential to your digestive processes and the other to produce insulin and various hormones essential to your body's metabolism. First, let's look at the digestive system.

Earlier, we talked about chyme, the thick, partly digested fluid that exits your stomach and enters your intestines. As it makes this transitory journey, it is highly acidic and would damage your small intestine if it were not quickly neutralized. The duodenum, that part of your intestine that immediately attaches to your stomach, detecting the need to reduce this acidity, secretes chemicals that cause your gallbladder to release alkaline bile and your pancreas to release sodium bicarbonate into your duodenum. These secretions effectively neutralize the chyme before it enters the more sensitive sections of your small intestine.

If these processes were not in place, several health-compromising events would quickly happen: your small intestine would be seriously damaged by the highly acidic chyme; the absence of pancreatic juices would inhibit the breakdown of proteins, starches and fats; and your body, unable to absorb these unprocessed nutrients through the intestinal walls, would starve to death. Once again, a fundamental question arises and we ask, "How would early mankind have survived the first generation had these secretions not been in place from day one?"

What about the diabetic implications? Most of us likely know at least one person who is a diabetic. My dad was one. He had high blood sugar (glucose) levels because his body rejected the insulin it produced. His blood sugars may also have been too high because his body didn't produce enough insulin. Either way, his body cells couldn't absorb glucose to turn it into energy.

Having this absorptive restriction makes your pancreas all the more important to you. You see, it has cells that produce insulin and glucagon, two hormones that regulate your concentration of blood glucose, a simple sugar that is an important energy source and a component of many carbohydrates. If your blood sugar levels get too high, insulin is released into your bloodstream to bring the sugar levels down. On the other hand, if your blood sugar levels drop too much, glucagon will trigger your sugars to increase to a safe level. When the pancreatic cells that produce insulin don't function properly, for whatever reason, diabetes occurs, and without modern drugs to manage the situation, death can follow. What if the earliest living creatures didn't have a pancreatic-like organ?

PITUITARY

Situated at the base of your brain, your pituitary gland is considered to be your body's master gland. And small wonder! It secretes hormones that regulate your body's internal environment to maintain a stable, constant condition. This little pea-sized half-gram powerhouse produces a number of hormones that help growth, blood pressure, the stimulation of uterine contractions during childbirth, breast milk production, sex organ functions in both men and women, the operation of the thyroid gland, the conversion of food into energy (metabolism), water regulation within the body, temperature regulation, the absorption of water into the kidneys, and more.

Do you see the emerging pattern that without each and every part of the body we have looked at, the success of life would have been dramatically compromised? We continue.

THYROID

Think of a bow tie and where it would be worn. Think about it weighing about 1 ounce (20 grams) or so. That's a pretty fair representation of what your thyroid looks like and where it's located. Of all the cells in your body, the cells in your thyroid gland are the only ones that can absorb iodine and convert it into hormones.

The thyroid hormones released into your bloodstream directly influence the rate at which your body's cells produce energy from the foods you eat: the more hormones, the more energy. Because of their energy management influence, your hormones affect a number of significant bodily functions, including bone growth, brain and nervous system development in children, mental alertness, body temperature, heart rate, and physical growth and cell reproduction.

When your thyroid doesn't function properly you feel cold and lethargic, and you are unable to control your weight. You frequently become ill. Your digestion and nutrient assimilation suffers because it becomes impossible to break down your food intake properly. The efficiencies of other body glands suffer too, and ultimately every bodily function becomes impaired. Never mind functioning properly—if your thyroid didn't work at all, or didn't exist from day one, would life have continued and flourished?

PINEAL

This small gland, pronounced *pin-ee-al*, is located in the middle of your brain and closely resembles a pine cone, hence its name, pine-al. Medical science has determined that the pineal has specialized nervous pathways that link it to the retina of your eye. Your pineal gland controls the release of melatonin—the hormone responsible for regulating your wake-sleep cycle—depending on the degree of light or dark in your surroundings. But that's not all that the pineal controls. Melatonin also prevents local inflammatory swelling, tissue damage and the buildup of protein-rich fluids around the lungs. It is an essential ingredient for the effective production of antibodies, regulating specific immune responses and scavenging free radicals that can compromise numerous cellular processes that contribute to the onset of a host of various diseases. Would you consider the pineal, then, as one of those "gotta-be-there-from-day-one-or-else" body parts?

SWEAT

A group of us sat in a sauna on the edge of a river in northern Canada several autumns ago. The pipe rising from the woodstove was so hot that it glowed red. It was so hot, I thought I would die. After what seemed like an eternity—actually only twenty minutes or so—our hosts opened the door, and we all made a mad dash to the dock and jumped into the frigid river.

Personally, I thought this was dumber than a bag of hammers, but not to offend my hosts, I did it twice more. To be honest, at first I thought I'd need a post-steam-bath psychiatric assessment, but to my huge surprise, I never felt better in my life. Why? The answer lies in the sweat.

Since earliest times, humans have known the hygienic and health benefits of a good sweat. Today, we recognize that the skin—the largest organ in the human body, with two to four million sweat glands—plays a key role in detoxification, and that regular sweating is beneficial to eliminating toxins from our bodies' systems.

Before the industrial revolution, physical exercise was found in every aspect of daily life. Today that's no longer the case, and we can see the effects of sedentary lifestyles wherever we look—maybe even in our own mirrors. When we don't sweat as much, we feel the worse for it. Without daily workouts, we get sluggish. Sweat truly plays an important role in our health. Without it, your body would definitely be susceptible to disease.

TESTICLES

Have you ever wondered if it was a random event of chance in the reproduction of the species that women had eggs in their ovaries that couldn't be fertilized without sperm and men had sperm produced in their testicles that were useless without an egg? Hmmm...

Obviously, testicles—also referred to as testes or gonads—produce sperm, and we'll talk a bit more about that soon, but did you know they are actually part of two separate body systems, the reproductive and the endocrine systems? The reproductive system is essential for the creation of new life, while the endocrine system manages a number of functions essential to keeping the body working well. The latter does this through the secretion of hormones produced within the body's various glands.

Under the control of the pituitary gland, the testicles produce, along with sperm, a number of very complex life-essential chemical substances called hormones, testosterone being the most important by far. Within the male these hormones play a myriad of roles: they regulate and are responsible for the development and changes in the sex organs as a boy travels through puberty, including the enlargement and lengthening of the penis; they direct the growth of body, pubic and facial hair; they govern the deepening of the voice and the widening of the shoulders; they stimulate muscle growth and the increase in bodily strength; and they are accountable for the overall management of puberty's growth spurts. Not only is testosterone important in physically transforming the boy into a man, it is also the primary chemical that makes him "feel" like a man during his puberty years. Important? Evolutionary? One in a million?

Here's a little gem the men will identify with. One testicle hangs a little lower than the other. Why? I don't know, but those medical types who study such things feel that it prevents one from hitting the other. As a man I can certainly appreciate that; there is not a more sickening feeling for a guy than to get a percussive tap of any magnitude down there. I wonder if this could be where the word "groan" originated?

A few years ago I went to an immensely educational exhibit called "Body Worlds."[8] The exhibit had many corpses that had been chemically treated in a process called "plastination," a technique used to preserve body parts by replacing water and fats with plastic-based compounds that stabilize the body structure. In one exhibit it was clear to see that each testicle was held in place by a long cord, called the spermatic cord, that extended high up in the abdomen. Men's testicles occupy a sack called the scrotum that sits outside the body, because sperm can't be made at body temperature. If a guy's surroundings get too cool, the muscles that control the spermatic cord will draw the testicles up closer to the abdomen to keep them warm, and if he begins to overheat, the reverse will happen and the testicles will be lowered to cool them down. This is the body's way of keeping the testicles at just the right temperature for sperm production. All this happens automatically without any thought from the man. Just in case you are wondering why the testicles operate at a lower-than-body temperature, I'm sorry; I haven't got a clue.

In our house we have a half dozen smoke and fire detectors. They are our backup system, completely redundant and, in my opinion, completely essential. Men have two testicles; women have two ovaries and two breasts; we all have two eyes, two ears, two lungs, two kidneys, two adrenal glands, two arms, two legs, etc. Have you ever thought about that duality? The reason? Backup? It would be nice to always have two in each category, but if one were damaged or diseased we could live with the other. Others would say the duality contributes to symmetry and balance. I'm OK with that, but I do question the original human "package" and whether or not it had one of everything. One day, somewhere in time, did it suddenly discover the need for redundancy? What necessitated that need? Maybe it started out with two of this and two of that from the very beginning. What do you think?

Turning our attention from the chemical-producing aspects of the testicles, let's look at their second main function, the production of sperm. These little swimmers (I call them that because to travel less than 1/2 inch (1.25 centimeters) their tails will thrust over a thousand times) are the smallest cells in the human body. They are produced in the testicles at the rate of about one thousand every second (that's eighty-six million every twenty-four hours, guys) and take around sixty-two days to grow and another twelve days to fully mature before being ready for active service.

Each sperm cell can be thought of as having three parts: a head, a body and a tail. The tail is obviously the propulsion mechanism, comprised of protein fibers that alternately contract on one side and then the other, creating a whipping-like motion that moves the sperm along, much like a swimmer wearing flippers. If this ability to swim hadn't been there from the beginning, how long do you think life would have lasted?

We can easily consider the head of the sperm as its chemical "brain," if you will. Every cell in the human body contains forty-six chromosomes, except for the woman's egg and the man's sperm; they only contain twenty-three chromosomes each. That's curious; have you ever wondered how that came to be? When the pointed head of a sperm finally penetrates an egg, the dense genetic material within the head unites with the twenty-three chromosomes in the egg, and the full compliment of forty-six chromosomes is then available to begin the cycle of new birth. Also curious, and without an appropriate evolutionary explanation, would you not agree?

During ejaculation, 100 to 400 million sperm head off to do their thing. They are carried on their journey by seminal fluid, whose total volume is about twenty times that of the actual sperm it is carrying. This fluid is made up of over thirty different components, including sugar, water, zinc, assorted salts, enzymes, ascorbic acid, cholesterol, nitrogen, citric acid, fructose, lactic acid, vitamin B12, vitamin C, proteins and other goodies. For those who like trivia, here's some more:

- During his lifetime, a man will produce 12 trillion sperm.
- If placed end to end, they would extend 32,000 miles (51,000 kilometers).
- Each sperm has a lifespan of 24 to 48 hours.
- The sperm's journey to the egg can take from 5 to 60 minutes.

Heads Up

From Your Shoulders Up

We've covered a lot of ground so far. Not wanting to miss anything, we now want to look at some of our physical senses and what lies just out of sight, so to speak. If for no other reason than the fact that you are reading this, we will start with the eyes and their accompanying structures; later on we will move to some of the other goodies that occupy the space above our shoulders.

EYES AND SEEING

Our eyes receive 80 to 90 percent of all the information we process; we are truly visual creatures. I'm sure you can vividly remember important events in your life that were delivered to you visually. I certainly can, especially the first time I saw the woman who would one day become my wife. When we first met, she was a teacher of blind and low-vision children. Everything she told me about her work I could identify with, or so I thought, even though I had no practical experience with what she did. Because I had never been unsighted, its reality was outside my paradigm.

Then all that changed; I put on a sleep mask—the kind you wear on a plane during a long flight—and tried to navigate my way around my cozy little house. For the several minutes I was voluntarily unable to see, I stumbled, bumped and timidly groped my way from one room to another. It was a little scary to contemplate a lifetime of visual impairment. From this experience I gained both a huge appreciation for my own sight and a sense of awe for how my body's vision system works. Try it for a half hour, and I guarantee it will change your appreciation for your sight!

A very cursory but thought-provoking look at some of our vision's components is next.

EYELIDS

That thin fold of skin that protects the eye is a marvel; it opens so we can see, and it closes to block out the light. Both of these actions can happen either voluntarily or involuntarily. It's main function, though, is to spread tears, from the ducts that surround the eye, across the eye's surface to keep the cornea—the transparent front part of the eye that covers the iris, pupil and anterior chamber—moist and lubricated. Without this constant moistening, the eyes would dry out, and sight would quickly end. The issue then becomes, did the eyelid evolve before or while the eye did? Or did they evolve at all?

TEARS AND TEAR DUCTS

None of us needs any introduction to tears. There are lots of reasons why we've cried: maybe it was pain, perhaps joy, maybe winning the lottery—or losing it. My least favorite source of tears was bugs in my eyes at dusk when I used to race along on my motorcycle—not fun!

Tears are produced by small glands located under your brow bone behind your upper eyelid, at the edge of your eye socket, and in your eyelids. These glands conduct tears through the tear ducts to your eyes to moisten them and wash away dust, dirt, germs and other irritating substances as the eyelids blink. Tears contain an antiseptic agent that helps prevent disease, and it is this fluid that constantly lubricates the cornea (the transparent coating of the eye), the sclera (the white part of the eye), and the inside lining of the eyelids. What do you think the ramifications would have been for early mankind had this agent not been there from the very beginning? Devastating!

Tears are also somewhat oily; if they weren't, they would rapidly evaporate, serving no purpose in the lubrication process. In fact, it would lead to early blindness. That puts us back to the question asked in the previous paragraph. Once again, hmmm...

When you tear a lot, your tear ducts—tiny holes located near your nose and in your eyelids—perform an important secondary function besides the washing and moistening of your eyes; they act as conduits to drain your tears into the back of your nasal passages. That's why you give your nose a good blow after a cry: it simply can't handle the overflow of tears.

EYELASHES

Have you ever noticed the size of the cosmetics and beauty industry that caters to eyelashes alone? It's huge! Women curl their eyelashes, extend them, thicken them, color them, and generally pamper them. And then, when the timing is just right, they bat them at their chosen male, and the process of flirtation is under way. Ah, vive l'amour!

But why eyelashes? What practical purpose do they perform? Here are some possibilities:

- Their extreme sensitivity to touch initiates a reflexive reaction causing your eyelid to close whenever any foreign object or debris comes near your eyes.
- They prevent perspiration from running down into your eyes.
- Their absence can be an indicator of an illness.
- And of course, they assist in communication.

OPTIC NERVE

When a fetus is developing in the womb, millions of optic nerve endings begin to emerge from the eye-buds. At the same time, millions of nerve endings also begin to emerge from the brain. Incredibly, these two sets of nerves find each other and join together perfectly. The result? Sight!

Did you get that? Go back and reread that last paragraph and then ask yourself, "What are the chances that all those millions of nerve cells would link with each other without knowing the pathways to those connections?"

Our eyes and brain divide what we see into a right and a left half. Our eyes invert the image, and the left side of what we see ends up in the right side of our brain, and the right side of what we see ends up in the left side of our brain. Sounds complicated, but it all works out because the right side of our brain controls the left side of our bodies and vice versa. Still sounds complicated!

YOUR EYE'S CONSTRUCTION

We've covered the surrounding structures of your eyes; now let's look at the marvel of the eye itself. Remember, this is a very high level description of one of the most complex organs in the entire body.

Say cheese! Have you ever considered that your eye is like a camera? Most likely someone at some point in time looked at the eye and decided to create its mechanical equivalent—a camera. Using the camera as the example, we'll start with the various components and then describe each in more detail.

In a camera, light enters through a circular opening called an aperture; your eye's equivalent is that black opening at the center of the eye, called the pupil.

The camera's iris diaphragm limits the amount of light entering through the aperture. Your eye's equivalent is the colored ring called the iris; its muscles automatically adjust the amount of light entering your eye. In extremely bright situations, like being out on a lake or river on a sunny day, your iris shrinks to limit the amount of light entering your eye. Conversely, when you are in an environment of extremely low light level, like on a cloudy day or in a dark room, your iris expands to let more light in. This is an exceptional capability; regardless of the lighting situation, your eye adjusts automatically to accommodate the intensity of light entering it.

Let me give you a personal example of what could happen if this were not the case. My ophthalmologist tells me I have the early stages of cataracts. When I have an appointment with him, one of his assistants will put special drops in each eye to cause my irises to open fully so that when he examines my eyes with his various instruments he can see everything clearly. The effect of these drops won't wear off for several hours, and when I leave his office and head out to the parking lot, even if on an overcast day, it's as if there's a searchlight shining in my eyes. When it's sunny, the intensity can be painful. Think what your sight would be like if you had no iris.

The eye's design is indeed wondrous. Contemplate that recurring question, "Did this capability come with the original human package, or did dear old Ma Nature take credit for adding it somewhere along the way?"

Let's continue. Light entering the camera passes through a lens, or system of lenses, and is focused on the film, or image sensor if it's a digital camera. Your eye is very similar in that light passes through its lens and is automatically focused on your retina, your eye's equivalent of the camera's film or image sensor. Assisting in this focusing capability is your cornea, the transparent exterior front part of your eye. In conjunction with the lens and the pupil, it provides nearly two-thirds of your eye's ability to bend and focus light as it enters.

Consider the ability of the healthy eye to automatically change focus rapidly on objects that are near and far. Consider also the ramifications to the human race if the primal body didn't have the ability to see. And if the eyes evolved, how is it that humankind survived until they became a part of the evolved package?

When you take a picture and want to see the results, it's either a case of developing the film or looking at the screen on the back of your digital camera. With your eye, the image formed on your retina is transferred along your optic nerve to your brain, where it is interpreted as sight. As I write this book I'm becoming more aware than ever that I fail so often to grasp the immensity of the wonders and the complexity of the human body; for example, what if we didn't have an optic nerve?

What an awesome structure we call our body!

The analog of the camera's body is your eye's sclera, the white opaque fibrous outer part of your eye. The front of the sclera, called the cornea, is transparent, allowing light to enter your eye. One might question if this transparency was there from the very beginning, or did it, too, develop along the evolutionary timeline? If it developed along the way, how did life survive, virtually blind, with this opaque covering in place for so many eons?

Between your corneal outer layer and the retinal inner layer lies a third layer, called the choroid, a layer of connective tissue about a half millimeter thick. This layer supplies both oxygen and nourishment to the outer layers of your retina. Without these essentials, your retina would cease to function, resulting in blindness. Was this choroid layer there in humans from the very beginning? If not, there would have been no sight. Furthermore, what marvel created both it and the retina at some point along the evolutionary timeline?

Your eye's interior, from the lens to the retina, is filled with a jelly-like substance called the vitreous humor, meaning a glassy-like bodily fluid. If it were not there, your eyeball would collapse. If it were not completely transparent, your vision would be compromised.

My automobile (and yours) needs occasional oil changes; without this essential lubrication the friction between the motor's moving parts would soon destroy the engine. The same holds true for the eye. Pause for just a moment to consider how hard your eyes work each day: the times they refocus, adjust for new lighting conditions, move, blink, etc. It has to number in the tens of thousands. Fortunately there is an automatic lube system in place to ensure that all the moving parts are adequately nourished. A clear fluid called the aqueous humor, literally meaning a watery body fluid, lubricates the pupil, the lens, the iris and the cornea and the spaces between them. If the production and absorption of the aqueous humor are not balanced within fine margins, pressure within the eye, either too much or too little, will result in serious vision impairment leading to blindness.

Once again we are obliged to ask whether this lubrication system came with the original human package, or did it somehow develop along the way? And if it did evolve, doesn't that suggest that until that point in time humans had no sight?

Before we leave the eye, there's one more part we dare not lose sight of—the retina. Your retina is the back part of your eye where light from what you are viewing gets focused. It is made up of two types of photoreceptor cells, called rods and cones. Collectively, these cells cover an area a little less than one square inch (about six square centimeters); they respond to impinging light and carry their resulting impulses along your optic nerve to your brain, where they are interpreted as sight.

Rods in each eye number about one hundred and twenty million. They don't perceive color or fine details but are quite effective at differentiating changes between light and dark, shape and movement. It's the rods that help you see in the dark, although upon entering a dark environment it will take seven to ten minutes for them to start to operate effectively. When you return to a bright environment the saturated rods will need a few minutes to return to normal and the cones will need a few minutes to begin to register light.

Cones, numbering around six million per eye, are less receptive to light than rods are but work beautifully in bright light. They are your eye's instruments for detecting fine details and for receiving light's three primary colors: red, green and blue. As they are not effective in the dark, you will not be able to see color in dimly lit environments.

Consider the structure of your retina. You can think of it as having two layers. The first layer that light hits as it enters your eye is nearly transparent and has no capability to interpret the light that impinges upon it. It is this layer, however, that carries the signals to your optic nerve from the rods and cones that make up the middle layer of the retina. The second is the choroid layer. Not only does the choroid nourish the retina's outer layers, it also completely absorbs any light that gets past the rods and cones. If it didn't and the rods and cones received both incident light and reflected light from the choroid, they would register that as more light than was really there.

Before we leave this section, here's something you may never have thought about: how is it that the brain and the muscles that control the shape of the lens on your eye know exactly how much to change that lens' shape to keep what you are looking at exactly in focus on the retina at the rear of your eye? It's totally automatic, but how and when did this intricate set of actions and reactions come about? As the saying goes, "Inquiring minds want to know!"

Just as an aside, consider what Darwin had to say about the eye: "To suppose that the eye with all its inimitable contrivances for adjusting the focus to different distances, for admitting different amounts of light, and for the correction of spherical and chromatic aberration, could have been formed by natural selection, seems, I freely confess, absurd in the highest degree."[9] It would appear that he, too, had misgivings about his theory.

Once again we have questions without answers. Were all the components of the eye there from the beginning? Was the ability to see in both dim and bright light an initial thought or an afterthought? If either of these questions received a "No" answer, what were the odds of the transition taking place successfully?

EARS AND HEARING

When I was a younger dad, the Concorde made its maiden landing at one of our local airports, and I took my older son and daughter to the airport to see it take off. We watched the magnificent white bird head towards us down the taxiway, turn, and position itself at the end of the runway, engines idling, waiting for the tower to give it permission to take off. We stood on the shoulder of a service road, only a few hundred feet from the plane. The noise of the engines was impressively *loud*, but hey, one of the reasons we were there was to *feel* the power of this feat of supersonic engineering.

When the tower gave the clearance to take off, the pilot moved the plane into position at the end of the runway, ready to depart. When he hit the throttles, the noise was unimaginable! *Deafening* would be too insipid a word. It all happened so fast. Our bodies were literally shaking from the vibrations.

My son was in front of me, and my daughter was in front of him. I shouted to my son to put his fingers into his sister's ears, and I stuck mine into his. A quick reply to a desperate need, to be sure. The only problem was, there were no fingers in my ears, and not for a second did I want to try to reallocate our fingers. But the damage had been done; only now can I appreciate the good hearing I once had.

Unlike your senses of vision, taste and smell, which are mostly chemical in nature, your ability to hear comes primarily through a combination of mechanical movements that create electrical impulses that your brain can understand and interpret. So let's consider this wonderful sense that can discriminate some sixteen hundred different frequencies.

Your ear has three distinct sections: the outer ear (sometimes called the external ear), the middle ear, and the inner ear. Our journey of discovery will work from the outside to the inside.

OUTER EAR

Your outer ear is made up of three parts. First is the part we all see and know; it's called the pinna or auricle. Then comes the second part, the auditory canal. At the end of the auditory canal is the third part, the eardrum (a.k.a. the tympanic membrane).

The pinna, the outer visible part with all its circular whorls, concentrates the sounds that come at you from your external environment and funnels them into your auditory canal. In this respect it is like the first stage in an amplifier.

Next is your auditory canal. Your external auditory canal is a hollow tube running about 1 inch (2.5 centimeters) to your eardrum. The outer third of this canal contains hair cells and wax-producing cells. Earwax plays two key roles in your hearing: it contains chemicals that fight off infections that could damage the skin in the canal, and it protects your eardrum by keeping it moist. The hairs further protect your eardrum by trapping dirt and other particles and preventing them from entering the canal.

Your eardrum is like the skin covering on a drum—a very thin membrane about 1/2 inch wide (less than 1 cm.), stretched across the inner end of your external auditory canal. Responding to sounds that travel down your auditory canal, it vibrates, and these vibrations, in turn, are transferred to the amplification capabilities of your middle ear. Your eardrum is composed of three differing layers of materials: the outer layer is skin, the middle layer is fibrous and elastic, and the inner layer is a mucous producing lining. A ring of a special type of cartilage surrounds the edge of the eardrum and holds it in place. I stagger to think this is a random structure; could all this possibly be the product of chance?

EARLOBES

Just before we move on, let's not forget to give honorable mention to that little set of flaps some of us have in greater abundance than others: earlobes. While they supposedly have no known biological function, here are three possibilities you might consider:

- Some believe the rich blood supply to the earlobes helps keep the surrounding tissue warm.
- Others believe they are there for the sole reason of enabling parents and teachers to guide naughty children to where they don't want to go.
- And of course, earrings, studs and all manner of hanging trinkets confirm that airport metal detectors are operating as they are supposed to.

MIDDLE EAR

Your middle ear has one primary function. and that is to convert the acoustic energy traveling down your auditory canal into fluid energy waves within your cochlea (pronounced *coke-lee-ah*), the spiral cavity of your inner ear. It does this with superb efficiency, too.

The inner side of your eardrum is attached to the first of three small interconnected bones contained within the middle ear; they have complicated names to remember, but most people know them as the anvil, hammer and stirrup, because that's what they look like. Your middle ear itself is an air-filled chamber, with a tube—the Eustachian tube—that links it to the back of your throat and nasal cavity to keep the air pressure equalized in this space with that of the surrounding atmospheric pressure. Without this equalization, your eardrum could rupture inward or outward, depending on the pressure imbalance. When you consider the micro-attachments of these three small bones and the "wisdom" of the Eustachian tube, without which the eardrum could be permanently disabled, one has to wonder at the mechanics of this miniature hearing marvel. What do you think—chance or design?

As the sound waves move from the eardrum through the three middle ear bones, the amplification in pressure increases approximately twenty-two times. The third of these tiny bones, the stirrup, connects to the bony labyrinth, which is your inner ear. It is here that these vibrations will be translated into impulses and transmitted to the brain for interpretation.

INNER EAR

To get some idea of the size of your inner ear, hold your hand up and look at your thumb; your entire inner ear is no bigger than its tip. Each ear is a mirror image of the other, not an exact duplicate. There's a left side, and there's a right side. Like so many other parts of the body, how did this come to be? This bony structure has two primary functions. The first is to manage your balance and equilibrium, functions that are associated with your inner ear's semicircular canals. The second is to manage your hearing, a process accomplished by your cochlea, aptly called your body's microphone.

First, let's consider balance. Your body's balance center is in three interconnected fluid-filled loops found in each ear. Each loop is oriented on a plane perpendicular to the other two. These are often called the ear's semicircular canals. Within each loop are minute hairs and tiny floating particles that move back and forth in response to fluctuations of the fluid caused by changes in your body's movement. The vibrations of these hairs result in the generation of impulses, which then travel along your auditory nerve to your brain, where they are interpreted. That construction makes it possible for you to concurrently detect both motions and accelerations that are up-down, left-right and forward-backward.

Without this unique capability from the beginning, early mankind would have had no balance and literally would have been unable to move without tumbling to the ground. Flight from predators or any kind of danger would have been impossible, since there would have been no feedback response to body position to keep it either vertical or in motion. Chances of survival would have been slim. Would you agree?

What about your hearing and the role your cochlea plays in that process? It works like this:

- The three bones of your middle ear react to pressure changes on your eardrum caused by sound waves entering the auditory canal of your outer ear.
- Your stirrup attaches directly to your cochlea and pushes or pulls on it, depending on whether the sound wave is pushing in on or pulling out from your eardrum.
- This motion causes corresponding pressure waves within your cochlear fluid, which stimulate tiny hair cells within its interior to vibrate.
- These vibrations are converted into impulses that your auditory nerve then transmits to your brain for interpretation.

The bottom line here: no cochlea, no hearing. And without the ability to hear, early humans would have had no auditory warning of potential danger and therefore no time to react to save their lives. Could such an essential structure happen either by accident or by evolution?

NOSE

When I was in university, one of my classmates used to make fun of me whenever I'd answer a question; he'd grin and laugh and say, "The nose knows!" Compared to others in my family with very nice noses, mine is a little (as in a wee bit) larger, though not quite a beak, just in case you were trying to see in your mind's eye what I look like.

As the years have passed, the most common complaint I've heard when someone is not happy with the shape of their body is that they wish they had a smaller, or prettier, or more handsome nose. In fact, nose jobs are the most common form of plastic surgery among both women and men. So, what intelligent comments can we make about the nose?

ORIENTATION

It's comical to think about what life would be like if the whole contraption were rotated 180 degrees upward. Your nostrils would point to the stars, perfect for collecting rainwater on a stormy day. Or this: I saw a cartoon once where a woman had washed her bra and hung it up on the line to dry. The birds found it and decided it would make a perfect foundation (no pun intended) for two nests, side by side. Just imagine what might want to nest in your upturned nostrils! Point made; enough said.

NASAL HAIRS

The hairs within your nose are stiff, unlike the hairs on your head. They act as a filter to prevent dust, insects, and other airborne contaminants from reaching your lungs. To help accomplish this feat, mucous membranes within your nasal cavity secrete a thick mucous fluid that helps to trap the dirt. The mucous in your nasal cavity also helps to warm and moisten the air you inhale, thus helping to prevent damage to your lung tissue.

SINUSES

Your sinuses are literally holes in your head. The tissue surfaces of these moist, mucous-lined cavities are coated with microscopic hairs called cilia. Always in motion, cilia are responsible for moving the mucous into your nasal cavity, where it can trap the various airborne particles, then move it to the back of your throat to be flushed to your

118

stomach. This important mechanism keeps harmful bacteria and substances from harming your body when breathing through your nose, which most of us do almost 100 percent of the time. Do you think this all came about with the first edition of humankind?

SMELL

At the top of your nasal cavity is your smell center. It is an area about the size of a quarter and has about ten million receptors that respond to the various odor molecules that enter your nose as you breathe. You may not know it (I didn't), but you can recognize about ten thousand uniquely different odors. When the receptors for a given odor are stimulated, signals are sent to your brain for interpretation, and somewhere in this process the smell you are sensing is linked to the event that is emitting the odor.

Think of the first time you smelled something burning and it turned out to be charred toast; the next time you smelled that same smell, the first thing you thought of was burnt toast, right? Would you consider this reaction an early essential for survival?

SUMMING UP

You realize that we've only covered a very cursory consideration of these astonishing faculties. Just imagine how much more complex it would have been were we able to look at the genetic processes that brought us all the way from two single cells—sperm and egg—to these fully functioning organs. If we tried to develop them today, with the technologies we have available, we would need a host of highly skilled engineers to even begin to contemplate the challenges.

There are chemical considerations, for example; trying to fully reproduce your sense of smell would be no small task. Your ears and their links to your brain would challenge electrical, mechanical, chemical and fluid dynamics engineers. Your eyes would test the ingenuity of optical physicists as well as chemical and electrical engineers. And so on. You get the picture, I'm sure. When you consider their intricacy and their thought-provoking sufficiency, how would you rate the chances of these structures happening by chance? Would it be at least a million to one against?

What Have We Got?

It's time to tally the probabilities, but first we need to ensure we each have a very rudimentary understanding of probabilities (and I do mean *very* rudimentary, so don't panic).

Consider this: if you flip a coin the odds of getting a head is one chance in two. If you flip the same coin again, the chance of getting a head is again one chance in two. No complex math here. Ah, but, what are the chances of flipping two heads in a row?

To get that answer we multiply the probabilities together, and we get one chance in four. Taking this one step further, the odds of getting three heads in a row would be one chance in eight. And so on. This is an important principle; when there are linked probabilities, you multiply the individual ones together to find the collective probability. So if we get an event where the chances of it happening are one chance in one thousand, and the chances of a second event linked to it are one chance in one hundred, then the combined chances of the two of them happening in sequence will be one chance in one hundred thousand.

In the introduction, I suggested that the chances of even just one of the body components we've discussed so far being evolutionary could be infinitesimally small, as in really, *really* small. I also said it would be you who would judge the validity of evolution based on your assessment of a very thin look at the wonders of the human body. What we've seen so far has been the ten-thousand-foot view of how we're made—very thin indeed, but that should be good enough.

Throughout the preceding chapters I have presented a lot of realities for you to ponder and respond to. There have been lots of questions and, on purpose, no answers. In fact, there have been over one hundred challenging questions, each without an answer.

I arbitrarily suggested that each of these provocative situations could represent a one in a million chance of being the result of evolution. I personally think this is an unrealistically low number for our purposes. I also said that you should feel free to choose whatever probability you thought appropriate.

Remembering what a million looks like (1,000,000), if we look at the probabilities of all those 100 events happening by accident, we get a number that says the chances of the reality of evolution are

1 chance in 1,000,000,000,000,000,000,000,000,000,
000,000,000,000,000,000,000,000,000,000,000,000,
000,000,000,000,000,000,000,000,000,000,000,000,
000,000,000,000,000,000,000,000,000,000,000,000,
000,000,000,000,000,000,000,000,000,000,000,000,
000,000,000,000,000,000,000,000,000,000,000,000,
000,000,000,000,000,000,000,000,000,000,000,000,
000,000,000,000,000,000,000,000,000,000,000,000,

000,000,000,000,000,000,000,000,000,000,000,000,
000,000,000,000,000,000,000,000,000,000,000,000,
000,000,000,000,000,000,000,000,000,000,000,000,
000,000,000,000,000,000,000,000,000,000,000,000,
000,000,000,000,000,000,000,000,000,000,000,000,
000,000,000,000,000,000,000,000,000,000,000,000,
000,000,000,000,000,000,000,000,000,000,000,000,
000,000,000,000,000,000,000,000,000,000,000,000,
000,000,000,000,000,000,000,000,000,000,000   (If
you count them, there are 600 zeros.)

Would such a minute probability in favor of evolution suggest that human life is an accident? Would it suggest that there had to be a grand design behind it? Remember, this was only a thin slice of reality. Can we let the evidence speak for itself?

You know, even if the probabilities had been downsized to one chance in ten for each of the one hundred questions asked, the math would still yield the probability of evolution as one chance in a number with one hundred zeros in it. Thought-provoking, isn't it?

When we consider the magnitude of this number, it's hard to conceive an appropriate frame of reference for it. What could possibly be anywhere near this small? I thought, surely it had to be something absurd, like the chance of finding one specific atom in the entire universe. Seemed reasonable, so I went to a reputable technical question-and-answer site[10] and probed the probabilities of finding one specific atom among all the atoms in the entire universe. In seconds I had an answer:

1 chance in 100,000,000,000,000,000,000,000,000,
000,000,000,000,000,000,000,000,000,000,000,
000,000,000,000,000,000 (Only 80 zeros—much less
than 600!)

Even that seemed significantly large when compared to
the probability of evolution happening by chance. And just
where do we go from here?

Read on.

For Thinkers Only

In the last few pages we saw numbers—probabilities—that were so staggeringly small that they gave no support whatsoever to the theory that humankind evolved by random chance. What would those numbers look like if we even began to take into account the myriad chemistries that develop and support its growth, reproduction and daily workings? Mind-boggling, I should think. Nor did we look at the processes that correctly sequence the growth of the various parts of the body as it develops and matures. And all that starts out from a dedicated little swimmer and an egg the size of the period at the end of this sentence. Were we to factor in those probabilities, the numbers would be more incomprehensively miniscule than they already are, not that we can comprehend the initial numbers anyway.

Before we continue, ponder this: would we have gotten similar results if we had looked at other life forms? Sobering thought, isn't it?

I titled this book *Against All Odds: For Thinkers Only*. I want to encourage you to focus on the subtitle, *For Thinkers Only*, and to ask yourself which of the following possibilities the facts best support: that humanity is the evidence of a grand well-planned design, if you will, or that it is the chance product of a genetic accident, the result of evolutionary processes. With the evidence so strongly against evolution, the only other option would suggest that we are here by design. That, then, would necessitate a designer, would it not?

The confirming factor should then be if we can find substantive evidence of such a designer, whoever or whatever he or she or it might be. As we investigate this possibility, we need absolutes to guide our search and on which to base our conclusions:

- There must be overwhelmingly substantive documented evidence at every step along the way.
- Multiple unbiased witnesses must have attested to specific historical events and persons.
- Any evidence considered must be broadly available, undeniable and without supported challenge.

RECOGNIZING THE AVAILABLE EVIDENCE

A fundamental question must first be asked: "How will we recognize the appropriate documented evidence when we see it?" Let me pose the answer to that question with an anecdote. Years ago a friend of mine, let's call him Steve, gave me the name and phone number of one of his business friends he thought I should meet. I'll call him Peter. Steve thought Peter and I could offer each other some interesting business opportunities. So I called Peter on his cell phone, and we scheduled a lunch meeting.

The patio restaurant of his choice was in a very busy part of the business district of the city in which I lived, and he gave me a detailed description of himself so I would recognize him easily—tall, average build, dark clothing, carrying a briefcase, etc. The problem was that almost every guy on the patio was tall, with average build, wearing dark clothing and carrying a briefcase. Not much help, Pete! What he could have added to his description was that he had on a bright red tie and was carrying an umbrella, his briefcase was really a black computer case, he wore glasses and had graying hair and a slightly receding hairline.

Here's my point—the more information we have about what or whom we are looking for, be it tomorrow's lunch date or the creator of the universe, the easier it will be to recognize him or her or it and to validate his or her authenticity.

We begin our hunt for this evidence with the greatest trove of recorded history planet Earth has to offer: the archeological records by and about the ancient Hebrew peoples—today's Jewish people. Their life has been captured in exquisite detail, not only in their own historical writings, but also in the annals of those peoples and nations that lived alongside of them, were defeated by them or, in some cases, subjugated them.

The Hebrew record of their history and their self-existent creator God, who calls himself "Yahweh," is recorded in the Old Testament books of the Bible. (From this point forward, I will refer to Yahweh as "God.") Here we find thirty-nine books written by approximately thirty authors[11] and covering almost four thousand years worth of historic events: dealings with their God, dealings with their kings and leaders, commentaries, records of wars, histories, instructions, laws, genealogies, people profiles, foretelling of future events, proverbs, wise sayings, laments, allegories, poetry and songs.

Many of the Old Testament writings point to a coming individual who the Old Testament prophets claimed would be a very special "messenger" from God. Throughout the New Testament we will find references to this same messenger. The references between the Old and New Testaments to this individual are too complementary to be ignored, so this will be the rational starting point for our search.

Evidence We Can Trust

Before we start examining the content, accuracy and validity of these historical documents, we first have to ask if they can be trusted. Could they have been manipulated over the ages to say what people wanted them to say? Valid question. The evidence available to us needs to be examined.

Consider this first of two hypothetical situations. Suppose several of my contemporaries and I witness a particular event together. Then I write a book chronicling that event. How long do you think it would take for my friends to detect flagrant, or even subtle, errors in what I wrote? Not long, I'm sure. And my reputation would be trashed immediately.

Now consider a second hypothetical situation: same contemporaries, same event, same book, but this time without errors, lots of copies of the book in print and widely distributed. Time passes. The eyewitnesses to these events and I, as the author, are all still alive, but someone has reprinted the book with erroneous changes to it. How long do you think it would take me, or those who were eyewitnesses to the original events, to challenge the errors? The same, I think, as in the first example. Not long. Let's see how this reasoning applies to any biblical documents we may want to consider: first the Old Testament and then the New.

OLD TESTAMENT VALIDITY

The Hebrew scribes of antiquity were painstakingly diligent about preserving the writings of their patriarchs and prophets. Sometimes they made new copies, and sometimes they made copies to replace existing copies that were too worn to continue using and had to be destroyed. Because the process of copying their scrolls was entirely by hand, they established a comprehensive set of rules to follow to ensure complete transcription accuracy. Here are but a few of those rules; break one and the page[12] would be destroyed immediately.

- If the scribe had specifically made an error in copying God's name, he would immediately destroy the entire page and start over.
- From original to copy, the scribe would copy letter by letter, never word by word.
- From original to copy, there had to be the same number of lines per page, the same number of words per line, the same number of letters per line, and the same number of letters per page.
- A second scribe would apply additional tests to each completed copied page to ensure it was an exact match with the original.
- A third scribe would then apply additional tests.
- Once the entire book was complete, a fourth scribe would apply still more tests, usually including a count of the phrases in the original and in the copy.

So important was it to preserve the integrity of their biblical documents that those responsible for the copies went to these extraordinary lengths to ensure total accuracy.

Before the discovery of the Dead Sea Scrolls, the earliest Old Testament manuscript in existence dated back to only 900 CE.[13] Then, in 1947, a Bedouin shepherd looking for lost goats in the area south of Qumran, several kilometers south of Jericho and on the northwestern shore of the Dead Sea, discovered what looked like a cave high up in the cliffs of the Judean Desert. Wondering if his goats had accidentally entered the cave, he threw a rock into the cave to see if the goats would respond. To his utter amazement he heard what sounded like pottery smashing from the percussion of the rock. His curiosity lead to the discovery of what we now call the Dead Sea Scrolls: 972 scrolls dating back some two thousand years. Further exploration revealed a total of eleven caves that collectively contained the oldest known copies of the Hebrew Scriptures, religious writings, commentaries and various artifacts.

The Old Testament documents found at Qumran proved to be one thousand years older than the oldest copies in existence at that time, dating back to almost 150 BCE.[14] The significance of these scrolls is that when they were compared to those dating from 900 CE, they were found to be identical. We can therefore be confident of the authenticity of these documents in our search.

Now, what about the significance and accuracy of the New Testament copies?

A LINK BETWEEN OLD AND NEW TESTAMENTS?

In or about the year 3 BCE, a Hebrew child was born in Bethlehem, a small town a short distance south of Jerusalem.

His Hebrew name was *Yeshua*, which, translated into English, is "Jesus." From his youth and into manhood, he grew up in the small northern Israeli town of Nazareth.

"So what?" we can ask. Lots of Hebrew kids of that era were called Yeshua. Of the hundreds, or thousands, of Yeshuas born in Israel in those days, how can we know this Yeshua is special? I believe we will discover the role that Jesus of Nazareth plays in fulfilling the messianic prophecies, thereby linking the Old and New Testament texts.

NEW TESTAMENT VALIDITY

We have already seen strong evidence to support the veracity of the Old Testament writings. Can we find similar evidence to support the validity of the New Testament scrolls? The answer would appear to be a very clear "Yes!" As of this writing there have been 5,600 handwritten New Testament manuscripts found in the Greek language and over 19,000 handwritten copies in the Aramaic, Syrian, Latin and Coptic languages. That's a huge base of documents to assess against each other for content accuracy and far more than comparable manually created copies of other important ancient writings that have survived the sands of time, including only seven originals of Plato's writings, forty-nine of Aristotle's, and six hundred and forty-three of Homer's.[15]

Consider, too, when the New Testament books were written; biblical scholars overwhelmingly agree that all were penned within one hundred years of the birth of Jesus.[16] This would mean that what was written about Jesus would have been available to those who had walked and talked with him every day—his "disciples"—as well as those who had seen and knew of him. If these writings were in any way errant, they would have been revealed as fakery and denounced almost immediately. This, in itself, should be sufficient evidence for our purposes that the validity of the New Testament writings can also be trusted.

Next, we will want to consider the integrity of a specific group of those who authored the Old Testament: the prophets.

PROPHETS AND PROPHECIES

Throughout the Old Testament era, God spoke through individuals of his choosing to convey to his people those things he wanted to say to them; the Bible calls them "prophets." On some occasions their message would be one of guidance, sometimes of instruction, and, more often than not, of rebuke. Now and then these conduits of God would prophesy frequently, and at other times, seldom. Their message could be interpreted two ways: one interpretation would apply to the current situation in which the people found themselves; another would be for a future time and circumstance. When they were not speaking God's word, they went about their everyday professions as farmers, shepherds, scribes[17] or priests.

These prophets of old had it a lot tougher than legendary prognosticators like Nostradamus[18] or any of today's modern so-called prophets. You see, God said that if they prophesied something that did not come to pass as they said it would, there was a major penalty:

"But any prophet who falsely claims to speak in my name or who speaks in the name of another god must die. But you may wonder, 'How will we know whether or not a prophecy is from the LORD?' If the prophet speaks in the LORD's name but his prediction does not happen or come

true, you will know that the LORD did not give that message" (Deuteronomy 18:20–22).[19]

That certainly raises the bar for the reliability and accuracy of any of their writings we may elect to use, doesn't it?

ASSESSING THE PROPHECIES

If we are going to be accurate in our deliberations, we don't want to accept them simply on this basis alone. Most of the prophecies we hear today rarely come to pass as stated. That makes most of us just a little skeptical of their accuracy. It certainly does me. Throughout history many events have been predicted: the fall of Rome, the rise of the Third Reich, this year's World Cup winner, tomorrow's weather and an endless series of warnings that the world was about to come to an end. All were predicted, and some—not all—happened. What we are looking for now are as many quantifiable numbers as possible upon which we can definitively say there was, or there was not, a grand designer of humanity. Where the numbers are absolutes, for example one chance in four, we will use those numbers; where we need to make an intelligent guesstimate, we will look at likely numbers and then downsize the probabilities to be very much on the conservative side. When I develop a probability as I see it and you disagree, use your number instead. Sound fair?

WHAT ARE WE LOOKING FOR?

Somewhere in the Hebrew literature we want to find evidence of or a claim that there is a creative force or creator. Whatever evidence we find, it will have to be substantiated beyond any shadow of doubt. At several places in the New Testament writings, we are presented with comments from its authors that would seriously suggest they knew who such a creator was. We start there.

JOHN

There are twenty-seven separate letters and documents that make up the New Testament, all written by men who had a personal relationship with one they passionately believed to be the creator: Jesus of Nazareth, or, as he is also referred to, Jesus Christ. This is the same Jesus today's Christians revere and unabashedly follow as the Son of God; it is also the same Jesus mocked and ridiculed for the past two thousand years by skeptics and non-believers. Talk about a dichotomy!

One of Jesus' closest disciples and companions was a man named John, who came to write the fourth book of the New Testament. In its opening verses, John, writing specifically about Jesus—whom he calls "the Word"—declares, "*In the beginning the Word already existed. The Word was with God, and the Word was God. He existed in the beginning with God. God created everything through him, and nothing was created except through him.*" (John 1:1–3). That's quite a claim: Jesus created all things! This is precisely the kind of declaration we want to explore, but only if the claim accurately fulfills the Old Testament prophecies about a creator.

Can we find similar evidence from other sources, both biblical and secular, about this Jesus? Yes, we can; first from Paul and Matthew, two of the New Testament's most revered authors, and then from a secular historian named Josephus.

PAUL

One of the early converts to Christianity was a radical Jewish zealot named Saul, a frenzied persecutor of the early Christians:

Meanwhile, Saul was uttering threats with every breath and was eager to kill the Lord's[20] followers. So he went to the high priest. He requested letters addressed to the synagogues in Damascus, asking for their cooperation in the arrest of any followers of the Way [i.e., followers of Jesus] he found there. He wanted to bring them—both men and women—back to Jerusalem in chains (Acts 9:1–2, emphasis added).

The Bible records that one day Saul had a personal encounter with Jesus, following his resurrection, and quickly became a believer:

As he was approaching Damascus on this mission, a light from heaven suddenly shone down around him. He fell to the ground and heard a voice saying to him, "Saul! Saul! Why are you persecuting me?" "Who are you, lord?" Saul asked. And the voice replied, "I am Jesus, the one you are persecuting! Now get up and go into the city, and you will be told what you must do." The men with Saul stood speechless, for they heard the sound of someone's voice but saw no one! Saul picked himself up off the ground, but when he opened his eyes he was blind. So his companions led him by the hand to Damascus. He remained there blind for three days and did not eat or drink. Now there was a believer in Damascus named Ananias. The Lord spoke to him in a vision, calling, "Ananias!" "Yes, Lord!" he replied. The Lord said, "Go over to Straight Street, to the house of Judas. When you get there, ask for a man from Tarsus named Saul. He is praying to me right now. I have shown him a vision of a man named Ananias coming in and laying hands on him so he can see again." "But Lord," exclaimed Ananias, "I've heard many people talk about the terrible things this man has done to the believers in Jerusalem! And he is authorized by the leading priests to arrest everyone who calls upon your name." But the Lord said, "Go, for Saul is my chosen instrument to take my message to the Gentiles and to kings, as well as to the people of Israel. And I will show him how much he must suffer for my name's sake." So Ananias went and found Saul. He laid his hands on him and said, "Brother Saul, the Lord Jesus, who appeared to you on the road, has sent me so that you might regain your sight and be filled with the Holy Spirit." Instantly something like scales fell from Saul's eyes, and he regained his sight. Then he got up and was baptized. Afterward he ate some food and regained his strength. Saul stayed with the believers in Damascus for a few days. And immediately he began preaching about Jesus in the synagogues, saying, "He is indeed the Son of God!" (Acts 9:3–20, emphasis added).

Following his miraculous conversion to Christianity, Saul was renamed *Paul* and became the most prolific of all New Testament authors, witnessing primarily to the non-Jewish world—Gentiles, if you will. He is traditionally referred to as the apostle Paul, but I'll just refer to him as Paul. In his letter of encouragement to the Christians in Colosse,[21] this once hater of Jesus and his followers made an incredible proclamation that Jesus is the creator of all things: "*Christ is the visible image of the invisible God. He existed before anything was created and is supreme over all creation, for through him God created everything in the heavenly realms and on earth*" (Colossians 1:15–16). Could this extremist-turned-believer have made such a declaration had he not believed with his whole heart and understanding that this was a statement of pure fact? History records that following his one-on-one encounter with Jesus, what Saul had been so avid of destroying became the very thing he, now as Paul, was willing to give his life for.

MATTHEW

In the Gospel of Matthew, we read about a very unusual event. Jesus and three of his disciples were together for a time alone, but as one of the disciples spoke a bright cloud over-shadowed them, and a voice from the cloud said, "*This is my dearly loved Son, who brings me great joy. Listen to him*" (Matthew 17:5). Here we see God referring to Jesus as his son.

Let's also look at the New Testament book called Hebrews. Its author is unknown, but it was written to all Jewish Christians in the early first century, including those who were having doubts about their beliefs and were considering a return to Judaism: "*Long ago God spoke many times and in many ways to our ancestors through the prophets. And now in these final days, he has spoken to us through his Son. God promised everything to the Son as an inheritance, and through the Son he created the universe*" (Hebrews 1:1–2). In these two verses, the author makes three intriguing statements we will use as links in our search: God's son is Jesus; Jesus is the creator of the universe; and there is common ground between what the prophets foretold and what Jesus was now saying.

We now have three independent biblical authors—John, Paul and Matthew—each with a personal relationship to Jesus, and each fully convinced that Jesus is the creator of heaven and earth. Valid testimonies.

What about secular corroborative evidence? For that we consider one Flavius Josephus, a first century CE historian.

While his name suggests that he was Roman, he definitely was not.

JOSEPHUS

Yosef Ben Matityahu, a Jew from Jerusalem, lived from 37 CE to approximately 100 CE, shortly after the earthly days of Jesus. During the war between the Romans and the Jews (66 to 70 CE), Flavius Vespasian—the same Vespasian who would later become emperor of Rome—captured Ben Matityahu and imprisoned him. Upon his release in 69 CE, Matityahu then traveled to Rome, where he took the name Flavius Josephus, after his captor.

When Vespasian became emperor, Josephus turned his attentions to documenting the history of the Jewish people. In one of his works, *Antiquities of the Jews*—written between 93 and 94 CE—he stood out as the only first century non-Christian writer, independent of the Gospels of Matthew, Mark, Luke and John, who made specific references to Jesus that support many of the biblically recorded realities of Jesus' life.

The following extract from his writings shows that he acknowledged Jesus as a miracle worker and a teacher, his trial and crucifixion, his resurrection three days later, and his likely fulfillment of the Old Testament messianic prophecies. Consider the similarity between the Gospel records and the historic evidence presented by Josephus. Although he never became a Christian, here's what he wrote about Jesus:

> At this time, there was a wise man called Jesus and his conduct was good and he was known to be virtuous. Many people among the Jews and the other nations became his disciples. Pilate[22] condemned

him to be crucified and to die, but those who had become his disciples did not abandon his discipleship; they reported that he had appeared to them three days after his crucifixion and that he was alive. Accordingly, he was perhaps the Messiah, concerning whom the prophets had reported wonders, and the tribe of the Christians, so named after him, has not disappeared to this day.[23]

Whoa! Look back a few lines. Did you catch the phrase "he was perhaps the Messiah"? We haven't seen that title specifically ascribed to Jesus before, and if Josephus used it, it had to be well known to the readers of his time, and therefore an important clue for us to consider seriously.

PROPHETIC PROCLAMATIONS

Let's look a little more closely at this new clue: Jesus as the Messiah. The Hebrews of long ago believed that God would one day send his emissary, his Messiah, to rule the world and usher them into a time of global peace and prosperity. "Why is this important?" you might ask. Ah, *Messiah* is the Hebrew word meaning "Anointed One," and its equivalent word in Greek is *Christ*. That means *Jesus Christ* in Greek is the same as *Yeshua Messiah* in Hebrew. That thickens the plot—Messiah and Jesus Christ almost in one breath, and Jesus and creator in another, as we saw earlier.

Wouldn't it be incredible if we could link the Old Testament messianic prophecies with Jesus of the New Testament? That would put us very close to confirming whether or not this same Jesus is the creator. The Old Testament writings will give us our next set of clues.

Of the close to three hundred messianic prophecies in the Old Testament, forty-eight are major prophecies, each with its own unique contribution to the tapestry identifying how to recognize Israel's coming Messiah. Were we to assess each of these prophecies, this little book would grow to encyclopedic proportions, so we will carefully scrutinize only the most important of the forty-eight.

Recall the story about my friend Peter and how I would have had no problem recognizing him had he been more precisely described. Is the Old Testament portrait of the Messiah so detailed that it could only point to one person in all of human history? Perhaps. In every aspect of his life, Jesus exhibited an uncanny alignment with these prophecies. Is there overwhelming evidence? We shall see.

WHO GETS TO HEAR GOD'S WORDS FIRST?

The prophet Amos declared, "*Indeed, the Sovereign LORD never does anything until he reveals his plans to his servants the prophets*" (Amos 3:7). In other words, the prophets get to hear God's intentions first, before anyone else does. That makes it very important for us to understand what these prophets had to say about God's Messiah.

The exactness with which Jesus of Nazareth supposedly satisfied the messianic prophecies has been said to be a staggering 100 percent by those biblical scholars who have studied them. It is because of this extraordinary predictive precision that it is felt that the Old and New Testament manuscripts hold significant insights that we cannot afford to ignore. However we progress from here, we want nothing less than indisputable 100 percent accuracy. Only documented facts will be considered, and no personal biases or conjectures will be allowed.

If our research were to produce evidence as incontestable as we saw in our study of evolution, it would more than support a confirmation that Jesus, Messiah and creator are one and the same, would it not? The evidence is mounting. Hang on! One way or the other, the ride will be fascinating, that's for sure.

THE THRONE OF DAVID

One of the chief prophecies that points to Jesus is this one:

For a child is born to us, a son is given to us. The government will rest on his shoulders. And he will be called: Wonderful Counselor, Mighty God, Everlasting Father, Prince of Peace. His government and its peace will never end. He will rule with fairness and justice from the throne of his ancestor David for all eternity (Isaiah 9:6–7).

This individual talked about in these verses must have a direct lineage back to David; that's going to be very important. While this prophecy doesn't say directly that this individual would be the Messiah, the Jews knew that was exactly who the prophecy was talking about.

If we look into Luke's Gospel in the New Testament, we read how the angel Gabriel was sent to Mary, who would be Jesus' mother, and said to her, "*You will conceive and give birth to a son, and you will name him Jesus. He will be very great and will be called the Son of the Most High. The Lord God will give him the throne of his ancestor David.*" (Luke 1:31-32). There's the reference to the throne of David again. Isaiah's prophecy meant that there would be no break in the descendants of Israel's King David until Israel's Messiah arrived on the scene; in other words, Israel's Messiah would have to be born into that lineage. Gabriel confirmed that the prophecy was about Jesus. If we cannot substantiate that Jesus would be born as a direct descendent of King David, that would be an instant show-stopper right there; after all, there must have been a lot of generations between David and Jesus.

GENEALOGICAL MANDATES

As the number of Hebrews multiplied, they became a vast people known as the nation of Israel—today, the Jewish people. Their second king was David, the son of Jesse. Isaiah, one of Israel's most prominent prophets, spoke of the coming Messiah as being a direct descendant of Jesse and said, "*Out of the stump of David's family will grow a shoot—yes, a new Branch bearing fruit from the old root. And the Spirit of the LORD will rest on him—the Spirit of wisdom and understanding, the Spirit of counsel and might, the Spirit of knowledge and the fear of the LORD*" (Isaiah 11:1–2).

Exactly who was it that the Spirit of the Lord was going to rest upon? We find the answer to that in the New Testament book of Luke. On the Sabbath, Jesus was in the synagogue in his hometown of Nazareth, and he had just been handed the scroll of the prophet Isaiah. He stood up and began to read the appointed Scripture for that Sabbath day where it says,

> "*The Spirit of the LORD is upon me, for he has anointed me to bring Good News to the poor. He has sent me to proclaim that captives will be released, that the blind will see, that the oppressed will be set free, and that the time of the LORD's favor has come.*" He rolled up the scroll, handed it back to the attendant, and sat down. All eyes in the synagogue looked at him intently. Then he began to speak to them. "*The Scripture you've just heard has been fulfilled this very day!*" (Luke 4:18–21).

Clearly he was proclaiming that he was their Messiah the prophecy spoke about.

OK, let's see if Jesus fulfilled the basic requirement of being a direct descendent of Israel's King David.

The Hebrews meticulously recorded their genealogies, and for that this study is indebted to them. It is interesting to note that these records never included women, only men. Whenever a family had no sons and a daughter who married, it was her husband's name that was entered into the genealogical record as the son of his wife's father. In other words, the groom was now considered to be as much his father-in-law's son as his son-in-law. Within Hebrew law, this was considered a perfectly allowed custom.

Let's see how this works with Joseph and Mary. Matthew records this about Joseph and his biological father: "*Jacob was the father of Joseph, the husband of Mary*" (Matthew 1:16). That's clear enough. Now, what about Mary? Well, that's not quite as clear, unless we consider what was discussed in the previous paragraph. Luke records this genealogical entry this way: "*Joseph was the son of Heli*" (Luke 3:23). In this verse, Joseph is referred to as Heli's son, yet he carries a second title, too, Heli's son-in-law. That's because Mary was Heli's daughter.

And how do we know for sure that this is the case? Four ways. First, there was no violation in how the genealogies were recorded for Jesus, Mary or Joseph. Second, Matthew's and Luke's Gospels recorded so much similar information, including genealogies, that it would have been impossible for one of them to make a mistake in the record that the other would not have challenged. Third, Luke had no doubt about Joseph's lineage: "*And because Joseph was a descendant of King David*" (Luke 2:4). And fourth, no one, not even among the most virulent of the early enemies of Christianity, ever doubted Jesus' lineage back to King David.

Both parents, then, were lawful descendants of King David: Joseph through David's son Solomon[25] (Matthew 1:1–16), thereby inheriting the legal right to the throne of David, and Mary through David's son Nathan[26] (Luke 3:23–31), her line carrying the actual DNA of David.

Therefore, we have evidence that Jesus met the conditions of lineage, both through Mary and through Joseph. What then are the probabilities associated with Jesus being the Messiah based solely on the lineage prophecies? From King David to Jesus via Joseph's line there are thirty-two generations,[27] and from King David to Jesus via Mary's line there are forty-two generations.[28] We also need a number to represent the average number of children in each generation that could propagate that generation forward.

The Old Testament records considerable detail about the children born at each generational level. Sometimes, as with King David, we don't know all the details about the number of children, but in his case we do know there were at least fourteen offspring by a number of wives. In other texts, a verse will often mention the most important sons or daughters and then add the phrase "*he had other sons and daughters*" (see various verses from Genesis 11:11–24). Just how many other sons or daughters we don't know, so this necessitates a bit of guesstimating.

Being herders and agrarian families with significantly greater life spans than we have today, it is likely that the numbers of children in each family were correspondingly large. However, I have used an average of only five children from each generation from which the next generation could be born. As always, select a different number if you wish.

So the probability of Jesus being born into Joseph's line would be one chance in five, times one chance in five, times one chance in five, etc., thirty-two times; and for Mary's line, one chance in five, times one chance in five, etc., forty-two times. I don't recommend you try to solve these numbers on a typical calculator; the results are rather large. Here they are:

The probability of being born into Joseph's line is approximately

1 chance in 20,000,000,000,000,000,000,000.

The probability of being born into Mary's line is approximately

1 chance in 200,000,000,000,000,000,000,000,000.

Not exactly trivial numbers, are they? The fact that Jesus fulfilled this prophecy is compelling evidence that he is, without a doubt, both Messiah and creator.

But let's not stop here. Remember, we earlier outlined three absolutes on which to base our conclusions: substantive and overwhelming documentation, unbiased witnesses, and broad and undeniable evidence. In spite of these preliminary numbers, we need to honor these absolutes.

We turn now to some of the major messianic prophecies to see if or how Jesus fulfilled them.

PLACE OF BIRTH

Just to add to the assurance that there would be no doubt in recognizing the Messiah when he came, the prophecies even spelled out the exact location and town where his birth would occur. The Old Testament Hebrew prophet Micah, writing somewhere between 735 and 700 BCE, made two prophecies tied directly to their predicted Messiah. Speaking for God, Micah said not only that the Messiah would be born of the tribe of Judah, but also specified the exact town in Judah, Bethlehem: *"But you, O Bethlehem Ephrathah,[29] are only a small village among all the people of Judah. Yet a ruler of Israel will come from you"* (Micah 5:2). Jesus' disciple, Matthew, would later write of him, *"Jesus was born in Bethlehem in Judea, during the reign of King Herod"* (Matthew 2:1). *Judea* is the Roman adaptation of *Judah*, the name of one of the twelve tribes of Israel; the Hebrew patriarch Jacob (later renamed *Israel* by God) had twelve sons, and each son became the leader of a tribe within the ancient nation of Israel. His son Judah led one of those tribes. So the probability of the Messiah being born into that specific tribe would be one chance in twelve.

If we skip forward in time to Jerusalem under the rule of Rome, we will find another witness, Luke.

Caesar Augustus had ordered that a census of all the inhabited earth be taken and that everyone had to return to the city of his or her birth to be registered. One of Jesus' contemporaries, a physician named Luke, likely a Greek, recorded the events that led up to Jesus' birth:

> *And because Joseph was a descendant of King David, he [as family head] had to go to Bethlehem in Judea, David's ancient home. He traveled there from the village of Nazareth in Galilee. He took with him Mary...And while they were there, the time came for her baby to be born. She gave birth to her first child, a son"* (Luke 2:4–7).

And Mary named him Jesus, just as the angel had told her several months earlier, *"You will name him Jesus"* (Luke 1:31).

Justin Martyr and Emperor Constantine of Rome were two nonbiblical personalities of the early Christian era who corroborated these events. We start with Constantine.

When Constantine was emperor of Rome (306 to 337 CE), he wanted to know where Jesus had been born. For the answer, he turned to Origen, one of the wisest of the early Church fathers. Origen told him that whenever he visited Bethlehem, even the pagans there, who had no interest in Christianity whatsoever, were still willing to tell anyone who would listen where the great Jesus, whom the Christians worshiped, was born: in Bethlehem.

Justin Martyr, another of the early Church fathers (103 to 165 CE), in his treatise titled *Dialogue with Trypho,*[30] wrote not only about the town in which Jesus was born but significantly about the environment as well:

But when Quirinius[31] was taking his first census in Judea, Joseph traveled from Nazareth, where he lived, to Bethlehem, to be enrolled, for he was by birth of the tribe of Judah, which inhabited that region. Then he was ordered [in a vision] to go with Mary into Egypt and to remain there with the child until another revelation should advise them to return into Judea. Now, concerning the birth of the child in Bethlehem, [you should know that] when Joseph could find no lodging place in the village, he went to a cave nearby, and there Mary gave birth to the child and laid him in a manger, and there the Arabian magi found him.

We now have multiple sources confirming that Jesus was born, according to the prophet Micah, in Bethlehem of Judea. Next, we must determine the probability of Jesus being born in Bethlehem and not some other town in Judea. In those days, there were at least fifty larger towns and over a hundred smaller ones in the area called Judea,[32] so we will set the chances of the Messiah being born in the prophesied town of Bethlehem conservatively to be one chance in fifty.

The overall probability then is one chance in twelve that he was born of the tribe of Judah, times one chance in fifty that he was born in Bethlehem, for a total of a one chance in six hundred that Jesus was born in Bethlehem of Judea. So far, he is beating the odds.

Three paragraphs back, a reference was made to magi journeying to Bethlehem, looking for the child Jesus. Who were these guys, anyway, and what led them to Jesus in the first place? Is there any validity in the story? What supporting evidence can we find?

THE MAGI (A.K.A. THE WISE MEN)

The magi were scholars of the Babylonians, Persians and Medes (known today as the Kurds). As wise and learned Gentiles, they were aware of the principles of astronomy; some were devoted to the pursuit of astrology and had gained an international reputation for their understanding of this area of study. Both the magi and the Jewish scholars of the time were keenly aware of biblical prophecies that related to stars. One prophecy in particular held their attention; in the Old Testament book of Numbers we read, "*I see him, but not here and now. I perceive him, but far in the distant future. A star will rise from Jacob; a scepter*[33] *will emerge from Israel*" (Numbers 24:17). This scepter was a clue to the magi; they knew that it symbolized the power of the ruler that possessed it, so they could easily conclude from the prophecy that a new king of Israel would some day enter the world stage. The key to the arrival of that new king would come with the appearance of a star in the sky.

And as you would surmise, such an astronomical event did take place; most people today refer to this as the "Star of Bethlehem." The magi immediately linked this event with the ancient Hebrew prophecy in Numbers and their understanding of astrological signs and came up with the conclusion that a very important person in Israel's history had just been born. Just how they concluded that the birth was that of the Jews' Messiah we will look at soon, but armed with this understanding, they set out to see this newborn king.

When they arrived in Bethlehem they sought out King Herod[34] and said to him, "*Where is the newborn king of the Jews? We saw his star as it rose, and we have come to worship him*" (Matthew 2:2). They knew the Messiah would be a Jew because the prophecy from Numbers 24:17 referred to Israel, but how did they know he would be a king? And as for the celestial body they followed, what was it? Was it perhaps something that the wise men saw but was invisible to everyone else?

There's more to consider. Several theories equate the "Star of Bethlehem" to astronomical events that happened in the heavens roughly in the timeframe scholars equate to the birth of Jesus. The accuracy of the theoretical answers depends on a number of conflicting assumptions that have been well challenged by various scholars and astronomers. As questionable as one theory's position might be against another's, were they to stand alone, there is a measure of plausibility in each. Let me integrate two or three very similar theories together and show you how intriguing the results can be.

It will make our pursuit of a reasonable answer easier if we first understand some of the important factors the magi understood all too well.

- They considered Jupiter, the largest of all the planets, to be the king planet.
- In their study of the heavens, they considered the star Regulus, in the constellation Leo—Leo being the constellation of kings—to be a regal star.
- They were likely fully aware that the lion was the symbol of the tribe of Judah; Genesis 49:9 says, "*Judah is a lion's whelp*"[35] (NASB).
- They regarded the constellation Virgo, meaning "virgin," to be a symbol of birth.
- They had frequently observed stars or planets appearing so close together in the sky that they appeared to the naked eye to be one and the same; this is called a "conjunction."
- They understood the prophecies of the Jews' coming Messiah.
- And finally, they would have been aware of the dates for the Jewish New Year—Rosh Hashanah.

There are three possibilities we could consider: it could have been a star, a planet, or a supernatural event specifically orchestrated by God to announce the arrival of his Messiah.

WAS IT A STAR?

A common theory put forth by Ernest Martin[36] makes sense only if we are able to peg the death of King Herod just right. You see, those who have researched the star anchor their theories on the year King Herod died. But herein lies a problem: some scholars have concluded that he died in 1 BCE, and others say it was 4 BCE. Martin supported the 1 BCE hypothesis and drew all his conclusions assuming this was the correct date. Based on this date, we get the following thesis for the Star of Bethlehem.

In 3 BCE, Jupiter, the king planet, appeared so close to Regulus that the two appeared in conjunction with each other. Remember that Regulus was considered a regal star and that it was the brightest star in the constellation of Leo, the lion. If we bundle all this up, here's what we have: king planet, regal star, lion (indicative of Judah). Martin then presents evidence of the sun being in the constellation Virgo at this same time. And just how would an astrologer read these signs? I like the way Dr. Ray Bohlin put it in his article "The Magi and the Star of Bethlehem"[37]: "The royal planet in conjunction with the royal star while the sun is in a virgin." If, the assumptions were all correct, that should have been enough to get the magi packed "real quick" for a long trip.

And what about the second option, that of a planet (or planets)?

WAS IT A PLANET?

Between 7 and 6 BCE, shortly before Jesus was born, a remarkable thing happened in the skies over Judea. Two very bright planets, Jupiter and Saturn, came into conjunction with each. Here's how we know that actually happened. In the early 1600s, Johannes Kepler, the great German mathematician, astronomer and astrologer, saw this phenomenon happening in the sky where Jupiter and Saturn were almost touching. He calculated that this rare phenomenon would happen next in another eight hundred years. Calculating backwards from Kepler's day, it also happened in approximately 0 CE, intriguingly close to the accepted date of Jesus' birth. It would have been the brightest thing to ever happen in the sky, appearing like a super star.

To the magi, the astrological message behind this would have been phenomenal. In astrology, Jupiter was the king's planet—the ruling planet and the largest light in the night sky—and Saturn, according to the Old Testament book of Amos, represented the shield and defender of Palestine: "*Was it to me [God] you were bringing sacrifices and offerings during the forty years in the wilderness, Israel? No, you served your pagan gods—Sakkuth your king god and Kaiwan your star god—the images you made for yourselves*" (Amos 5:25–26). In this verse God refers to two idols Israel had made for themselves: Sakkuth (their king) and Kaiwan (Saturn, their defender). So when Jupiter came to Saturn, the astrologers would have interpreted that as "A king is coming to Palestine." The messianic scholars and the magi in Babylon were both aware of the biblical prophecy and correlated it with the events in the sky. That could have been another plausible explanation for why the magi had set out for Palestine. However, it would appear that there just might be third, and much more likely, explanation.

WHOSE STAR WAS IT?

Up until now, we have been looking at astrological possibilities for the star. As interesting as the ideas on this subject are, there is another point of view that demands attention: it was God's star!

Why would God use astrological signs to announce the coming of his Messiah when he has many times condemned astrology and those who practice it? Consider his denunciation of these practices that Israel had been following: *"All the advice you [Israel] receive has made you tired. Where are all your astrologers, those stargazers who make predictions each month? Let them stand up and save you from what the future holds. But they are like straw burning in a fire; they cannot save themselves from the flame"* (Isaiah 47:13–14). If not an astrological sign, what then?

God may well have chosen to announce the arrival of his Messiah, Jesus, in a way that was most characteristic of him: by a supernatural light show perhaps, in this case an exceedingly brilliant star. It's one thing to make the stars and the planets appear in overlapping proximity at a specific time and place, but it's quite another to produce a supernatural event. Then, there's another issue that has no answer: was it just the magi and the shepherds who saw the star, or were there others too? Maybe we need to add this to our Questions Without Answers file.

AND THE PROBABILITIES ARE?

How, then, shall we assign a probability to the Star of Bethlehem?

If we regard the first possibility presented—was it a star?—we must then consider the complexities of the event. The right planet would have to perfectly align with the right star, in the right constellation, with the sun in perfect position within another specific constellation, and all this would have to occur above the horizon in the Middle East so it could be seen on exactly the right day and the right year. My intellect wants to say this is, like, one chance in trillions. Hold onto that number for a minute.

What about the second possibility—was it a planet? What if the star had not been a star at all, but the same two conjunctive planets that Kepler had documented: Jupiter and Saturn? Once again, we get the same kind of "co-incidental" happenings as in the previous situation; however, if we really wanted to make it very easy, we could say, as Kepler predicted, that this occurred only once every eight hundred years? That would give us one chance in eight hundred.

And that leaves us with possibility number three—whose star was it? Even if this was what really happened, I simply cannot fathom putting a probability against something God purposed to do; it would have to be a sure thing.

The only one of these three options with numbers I can get my head around is Kepler's two-planet scenario, so I set the odds at one chance in eight hundred.

SPEAK FOR YOURSELF

Let's move forward in time from the birth, the star and the magi. Little is recorded about Jesus until around the age of thirty; however, when he burst upon the scene, his notoriety grew rapidly. He did miraculous things like no one had ever done before, and he spoke like no one had ever spoken before. If ever a man called a spade a spade, he did. The religious establishment of his day hated him and sought to destroy him at every turn. Why? Because of the things he said.

A fundamental legal principle of law says that the defendant is always allowed to testify on his or her own behalf. We start by looking at the claims Jesus made about himself.

The New Testament records that Jesus drew crowds wherever he went. One day, while teaching a great throng of people about the Kingdom of Heaven, he clearly acknowledged his stand on an issue the religious leaders wanted to catch him on: breaking the laws that God had put in place. In response to their verbal trap, he replied, *"Don't misunderstand why I have come. I did not come to abolish the law of Moses or the writings of the prophets. No, I came to accomplish their purpose"* (Matthew 5:17). What a staggering comment! He used this opportunity to proclaim to the masses that he, himself, would accomplish everything the prophets had spoken of concerning their Messiah, thereby claiming the title of Messiah as his.

On another occasion, Jesus had just healed an invalid on the Sabbath day.[38] The devout Jews, who would have considered this healing an act of work, took great offence at this miracle. Jesus responded to them, saying, *"You search the Scriptures because you think they give you eternal life. But the Scriptures point to me!"* (John 5:39). Jesus had to be referring to the Old Testament Scriptures because, at the time of this discourse, there were as yet no New Testament writings. Here again, he was clearly saying that the prophecies were all about him.

We should consider one more event where Jesus listed his credentials. Here's the scene. He had been crucified and placed in a rich man's tomb, yet three days later his followers had found his tomb, which had been sealed and guarded by the Romans, open and empty. All Jerusalem was abuzz with the news. Shortly after this event, two men were walking along the road from Jerusalem to Emmaus[39] discussing these staggering events when Jesus caught up with them and began to ask them what they were talking about. Not recognizing him at first, they jabbered on. As they walked, Jesus, without revealing who he was, began to explain all the prophecies and promises about Messiah and how they had been fulfilled in him: "*Then Jesus took them through the writings of Moses and all the prophets, explaining from all the Scriptures the things concerning himself*" (Luke 24:27).

Earlier, I mentioned that there were forty-eight major Old Testament messianic prophecies; the word "*all*" in this last Scripture suggests that Jesus explained the entire forty-eight to his friends. Not even when he had finished did they realize who he was—until later that day, and then, wham! They got it.

These are just three events among many where Jesus testified to fulfilling the prophecies about Israel's Messiah. In my research I never found any evidence, either documented during his lifetime or at any time up to the present day, that his claim to be the Messiah had been successfully challenged. I'd love to be able to put a probability to this, but won't. You do so, if you prefer.

As Jesus' earthly ministry was approaching its climax, a number of important prophecies came into effect.

FINAL ENTRY INTO JERUSALEM

Jerusalem had never been a stranger to Jesus; he had been there many times, even in his childhood. But now we want to see what messianic evidence we can glean from what was to be his final entry into the city. We start with two linked prophecies that we will look at separately: the "how" of Jesus' entry in the days before his crucifixion, and the timing of his entry according to prophecies made by the Old Testament prophet, Daniel.

First, we consider the "how." Five hundred years before Jesus, there lived a Hebrew prophet named Zechariah. Chapters 7 through 14 of the Old Testament book of

Zechariah are generally considered by modern-day biblical scholars to be prophecies that foretold the coming of the Hebrew's Messiah. In Zechariah's writings we find a prophetic statement that Israel's (i.e., the Hebrews', or Jews', if you will) king would enter Jerusalem riding on a young donkey: "*Rejoice, O people of Zion!*[40] *Shout in triumph, O people of Jerusalem! Look, your king is coming to you. He is righteous and victorious, yet he is humble, riding on a donkey—riding on a donkey's colt*" (Zechariah 9:9).

Four of Jesus' closest disciples and biographers— Matthew, Mark, Luke and John—would later chronicle the events that transpired on the day Jesus made his final entry into Jerusalem. We will look only at what Matthew wrote:

> *As Jesus and the disciples approached Jerusalem, they came to the town of Bethphage*[41] *on the Mount of Olives. Jesus sent two of them on ahead. "Go into the village over there,"* he said. *"As soon as you enter it, you will see a donkey tied there, with its colt beside it. Untie them and bring them to me. If anyone asks what you are doing, just say, 'The Lord needs them,' and he will immediately let you take them." This took place to fulfill the prophecy that said, "Tell the people of Jerusalem, 'Look; your King is coming to you. He is humble, riding on a donkey—riding on a donkey's colt'"* (Matthew 21:1–5; see also Mark 11:1–11; Luke 19:28–44; John 12:12–19).

The disciples went and did just as Jesus had instructed them to and brought the donkey and its colt for him to ride on. Most of the crowd spread their coats on the road while others cut branches from the trees and spread them in their path. The crowds going ahead of him and those who followed were shouting, "*Praise God for the Son of David!*

Blessings on the one who comes in the name of the LORD! Praise God in highest heaven!" (Matthew 21:9). *"The entire city of Jerusalem was in an uproar as he entered. 'Who is this?' they asked. And the crowds replied, 'It's Jesus, the prophet from Nazareth in Galilee'"* (Matthew 21:10–11). It's interesting that their shouts were taken from an Old Testament psalm considered by many to be a psalm describing their coming Messiah: *"Bless the one who comes in the name of the LORD"* (Psalm 118:26).

Associating any kind of a probability to this event is far from straightforward. First of all, Zechariah said their king would enter Jerusalem riding on a donkey. Kings didn't usually ride donkeys to their coronations; they rode horses, or rode in chariots, and always arrived in great splendor.

Second, he said their Messiah would be *"righteous and victorious"* (Zechariah 9:9). Israel was looking for a conquering king who would save them from their enemies; in the time of Jesus that meant they wanted a warrior Messiah to rout the Romans who were occupying their land. Little did they know at the time that the salvation their Messiah would bring them would be very different from what they expected; the salvation he offered would release them from their sins and present them blameless before their God.

How shall we, then, assign a probability? If we consider all the rulers and would-be rulers to that date who made the processional to their coronation riding on a donkey, I should think that one chance in a hundred would be a fair probability. If you aren't comfortable with that, by all means, use your own guesstimate.

Next, we consider the timing of the Messiah's entry into Jerusalem. Old Testament prophecies pinpointed exactly when the Messiah's arrival into Jerusalem would be. Talk about no reason for Israel not being prepared to recognize their coming king! The analysis of these detailed prophecies depended on the interpretation of historic dates in Israel's past that not all the analysts fully agreed upon. Be that as it may, the variance of the predicted dates was so small as to be negligible, and that makes the prophecy worthy of consideration. We start in the book of the prophet Daniel.

Daniel had been praying when an angel appeared to him, saying that he had come to give him "*insight and understanding*" (Daniel 9:22). The angel then said to him, "*Now listen and understand! Seven sets of seven plus sixty-two sets of seven will pass from the time the command is given to rebuild Jerusalem until a ruler—the Anointed One [meaning Messiah]— comes*" (Daniel 9:25). When Daniel heard the angel say "*sets of seven,*" he knew the angel was talking about weeks. That makes sixty-nine weeks in all. Hang on to that.

Years before Daniel had received this prophetic word, Israel's sins had incited God to lead Israel into captivity for seventy years at the hands of the fierce Babylonians. That prophecy is found in the Old Testament book of Jeremiah:

> "*And now the LORD of Heaven's Armies says: Because you have not listened to me, I will gather together all the armies of the north under King Nebuchadnezzar of Babylon, whom I have appointed as my deputy. I will bring them all against this land and its people and against the surrounding nations... Israel and her neighboring lands will serve the king of Babylon for seventy years*" (Jeremiah 25:8–11).

And, as prophesied, it came to pass.

As Israel was being led away from their homeland, the Babylonians razed Jerusalem to the ground, utterly destroying it. In due time, Persia's King Artaxerxes conquered the Babylonian empire and ruled over all the lands Babylon had earlier subjugated, including Israel. Artaxerxes then granted Israel permission to return to Jerusalem and to rebuild it. It is against this backdrop that Daniel's prophecy of the sixty-nine weeks was given.

To understand Daniel's sixty-nine, we must first appreciate that the Israelites knew that this idiom regarding time did not refer to actual weeks but to years of weeks, meaning that the prophetic week equaled a calendar year. That would make 483 years (483 years = 69 weeks x 7 years/week) from the time of Artaxerxes' proclamation to rebuild Jerusalem to Jesus' triumphant entry into Jerusalem.

We also must be aware that ancient Israel reckoned time by 360-day years,[42] not like our year of 365.25 days. So we need to do a conversion. Multiplying 483 years by 360 days per year gives us 173,880 days; then dividing that by 365.25 gives us 476 years as we reckon years today. If Daniel's prophecy is true, then we should expect Messiah's arrival 476 of our years after the edict to rebuild Jerusalem had been given by King Artaxerxes. Although Jesus had been to Jerusalem often, this specific event refers his last entry, just prior to his arrest and crucifixion.

Next, we need to look for when that timeline—Artaxerxes' declaration to rebuild Jerusalem—began and see how it compares with Jesus' actual entry into Jerusalem.

Bible scholar J. D. Wilson contends that Artaxerxes' decree was made to Nehemiah on the first day of the Hebrew month Nisan in the year 444 BCE.[43] That would correspond to March 5, 444 BCE, according to today's Western calendar.[44] When the calendars were being switched from BCE to CE, the year 1 BCE was immediately followed by the year 1 CE, i.e., there is no year zero between the two systems. We have to subtract 1 from 444 BCE and start our calculation from 443 BCE. Applying the 476 years to March 5, 443 BCE, we arrive at 33 CE, a date most historians agree coincides with Jesus' actual last entry into Jerusalem on a donkey's colt.

If Jesus really was the Messiah, then he uncannily arrived right when the prophecies said he would. I should think the chances of him arriving when the prophecies said would be exceedingly small, but I will use one chance in a thousand.

Next, we will be looking at how visible yet transparent Jesus was to the people of his time; those who believed in him knew him by sight, yet the religious establishment hated him but had trouble identifying him in a crowd.

NO STANDOUT IN A CROWD

There are countless paintings of what people have perceived Jesus to look like. But what do the Old and New Testaments have to say, and can they confirm each other? The prophet Isaiah described the Messiah, saying, "*My servant [Messiah] grew up in the LORD's presence like a tender green shoot, like a root in dry ground. There was nothing beautiful or majestic about his appearance, nothing to attract us to him*" (Isaiah 53:2). In other words, he would look just like any other guy: no hunk, no piercing eyes, and definitely not a standout in a crowd—hardly the conquering deliverer Israel was expecting.

What does the New Testament have to say? In Luke's Gospel we see a compelling story that would support this. As an adult, Jesus returned to Nazareth, where he grew up, went into the synagogue and was handed the scroll of the prophet

Isaiah. He began to read the section that speaks of what the Messiah would do when he came. Here's Luke's account:

> When he came to the village of Nazareth, his boyhood home, he went as usual to the synagogue on the Sabbath and stood up to read the Scriptures. The scroll of Isaiah the prophet was handed to him. He unrolled the scroll and found the place where this was written: "The Spirit of the LORD is upon me, for he has anointed me to bring Good News to the poor. He has sent me to proclaim that captives will be released, that the blind will see, that the oppressed will be set free, and that the time of the Lord's favor has come." He rolled up the scroll, handed it back to the attendant, and sat down. All eyes in the synagogue looked at him intently. Then he began to speak to them. "The Scripture you've just heard has been fulfilled this very day!" (Luke 4:16–21).

This seemingly arrogant, blasphemous proclamation incensed those in the synagogue; to them, Jesus had just proclaimed himself to be their Messiah. Immediately they got up and drove him out of town: "*When they heard this, the people in the synagogue were furious. Jumping up, they mobbed him and forced him to the edge of the hill on which the town was built. They intended to push him over the cliff, but he passed right through the crowd and went on his way*" (Luke 4:28–30).

Picture the scene: their rage boiled over at Jesus' claim to be God's Messiah, and they led him out of the synagogue—not politely, I suspect—to the brow of a cliff, with the likely intent of murdering him. But by the time they got there, he couldn't be picked out of the throng. There was nothing in his appearance that set him apart. Isaiah's prophecy proved accurate in this situation.

Can we conclude that Jesus could have fulfilled Isaiah's prophecy? Perhaps. Setting a probability here is tough, so I would cautiously choose to be on the low side. Because he eluded those in the synagogue, likely numbering around fifty, I give it odds of one chance in fifty.

There's one more event we should consider. It was the night that Jesus was betrayed by Judas, one of his twelve disciples. Judas had made a deal with the Jewish chief priests that, for a price, he would point Jesus out to the Roman authorities so they could arrest him. As history records it, the Jewish religious establishment hated Jesus and his popularity with the masses because he was a huge threat to their base of power and influence among the people. The story unfolds in the Gospel of John:

> *And even as Jesus said this, Judas, one of the twelve disciples, arrived with a crowd of men armed with swords and clubs. They had been sent by the leading priests and elders of the people. The traitor, Judas, had given them a prearranged signal: "You will know which one to arrest when I greet him with a kiss." So Judas came straight to Jesus. "Greetings, Rabbi!" he exclaimed and gave him the kiss. Jesus said, "My friend, go ahead and do what you have come for." Then the others grabbed Jesus and arrested him* (Matthew 26:47–50).

Incredibly, the chief priests—those who had scrutinized his every move for months—did not even describe Jesus to the mob they sent to arrest him; they needed one of his followers to point him out. The chances here would be one chance in twelve because Jesus was alone with his disciples, save Judas.

BETRAYAL'S PRICE

This same issue of Judas' deal to betray Jesus also has a prophetic background. Five hundred years before Jesus, the prophet Zechariah painted a curious distinctive of the coming Messiah, saying that his worth would someday be valued at thirty pieces of silver:

> *And I said to them, "If you like, give me my wages, whatever I am worth; but only if you want to." So they counted out for my wages thirty pieces of silver. And the LORD said to me, "Throw it to the potter"—this magnificent sum at which they valued me! So I took the thirty coins and threw them to the potter in the Temple of the LORD* (Zechariah 11:12–13).

Skipping back to Judas, we read about the significance of the thirty pieces of silver. In Matthew's Gospel we read, "*Then Judas Iscariot, one of the twelve disciples, went to the leading priests and asked, 'How much will you pay me to betray Jesus to you?' And they gave him thirty pieces of silver. From that time on, Judas began looking for an opportunity to betray Jesus*" (Matthew 26:14–16). There are two possible ways to look at this fulfillment of Zechariah's prophecy; Judas may have been completely familiar with this specific prophecy and was simply realizing it, or he may not have had a clue that he was prophecy's pawn in this situation. My vote would be for the latter, and I would set the probability at one chance in a thousand, although I think it really is much, much smaller.

There's more.

Before we leave this particular prophecy, there's an uncanny similarity that should not be ignored; thirty pieces of silver were paid to Judas basically to set Jesus up for his death, but little did Judas know that these same thirty pieces of silver would set him up for his own death. The details of the prophecy and its fulfillment are too identical to be ignored. After the betrayal, Judas had great misgivings about what he had done:

> *Very early in the morning the leading priests and the elders of the people met again to lay plans for putting Jesus to death. Then they bound him, led him away, and took him to Pilate, the Roman governor. When Judas, who had betrayed him, realized that Jesus had been condemned to die, he was filled with remorse. So he took the thirty pieces of silver back to the leading priests and the elders. "I have sinned," he declared, "for I have betrayed an innocent man." "What do we care?" they retorted. "That's your problem." Then Judas threw the silver coins down in the Temple and went out and hanged himself. The leading priests picked up the coins. "It wouldn't be right to put this money in the Temple treasury," they said, "since it was payment for murder." After some discussion they finally decided to buy the potter's field,*[45] *and they made it into a cemetery for foreigners* (Matthew 27:1–7).

We need to decide what the probability is of Judas trying to return the coins out of remorse and the priests using the money to purchase the potter's field as Judas' burial place following his suicide, exactly as Zechariah's prophecy five hundred years earlier foretold. I think the chances would be very small, like one chance in a thousand, or less.

For this two-part prophecy, we now have a combined probability of one chance in a million.

FAIR WEATHER FRIENDS

Have you ever been in a situation where those closest to you abandoned you when the going got tough? I think it happens to all of us sooner or later. The same thing was prophesied to happen to Israel's Messiah. Once again we take our reference from the prophet Zechariah, where we read, *"'Awake, O sword, against my shepherd, the man who is my partner,' says the LORD of Heaven's Armies. 'Strike down the shepherd, and the sheep will be scattered'"* (Zechariah 13:7). What exactly is this verse saying? It's a declaration from God that the one he has chosen to be Israel's shepherd (Messiah) would be murdered and, when he was, his disciples would scatter.

When any leader is struck down it would be a normal course of events for his or her followers to scatter, maybe in fear that the same fate awaits them, or perhaps to take time to regroup. The remarkable pronouncement here is that God, their all-powerful, all-knowing God, was going to allow the sword to come against his own emissary. And that's exactly what happened with respect to Jesus.

In the book of Matthew we see the record of Jesus' last meal with his disciples before his betrayal by Judas. It would have been important to Jesus to tell them what was about to happen so they would be prepared. In Matthew we read, *"Jesus told them, 'Tonight all of you will desert me. For the Scriptures say, "God will strike the shepherd, and the sheep of the flock will be scattered"'"* (Matthew 26:31). Hours later, as they walked on the hillsides of Jerusalem, Zechariah's prophesy unfolded exactly as Jesus said it would. Judas appeared and identified Jesus, the Romans led Jesus away, cruel torture and death awaited him, and his disciples abandoned him, scattered and hid. Their fear of being associated with Jesus was so great that at least one of them was willing to denounce any relationship to him just to save his own skin.

I suppose one of life's more devastating little events comes when you have made a solemn oath to someone that they immediately challenge, and then, much to your chagrin, their challenge quickly comes true. Ouch! Following the quote from Matthew in the previous paragraph, Peter, one of Jesus' closest companions, swore to Jesus that he would never abandon him:

> *Peter declared, "Even if everyone else deserts you, I will never desert you." Jesus replied, "I tell you the truth, Peter—this very night, before the rooster crows, you will deny three times that you even know me." "No!" Peter insisted. "Even if I have to die with you, I will never deny you!"* (Matthew 26:33–35).

Yet, immediately after Jesus was betrayed, three times Peter flat out denied knowing him. Here's the rest of the event. After the arrest, Peter had followed the crowd to the courtyard of the high priest's house and sat down with the officers to see what the outcome of Jesus' arrest would be. We continue to read what Matthew recorded:

> *Now Peter was sitting outside in the courtyard, and a servant-girl came to him and said, "You too were with Jesus the Galilean." But he denied it before them all, saying, "I do not know what you are talking about." When he had gone out to the gateway, another servant-girl saw him and said to those who were there, "This man was with Jesus of Nazareth." And again he denied it with an oath, "I do not know the man." A little later the bystanders came up and said to Peter, "Surely you too are one of them; for even the way you talk gives you away." Then he began to curse and*

swear, "I do not know the man!" And immediately a rooster crowed. And Peter remembered the word which Jesus had said, " Before a rooster crows, you will deny Me three times." And he went out and wept bitterly (Matthew 26:69–75 NASB).

Talk about a prophecy that comes to pass within hours! Setting a suitable probability around these events is complicated because we really want to consider the probability that God himself would proclaim that his Messiah would be abandoned by his followers and it wouldn't happen exactly as it had been foretold. That would be akin to saying that God didn't know what he was talking about or that he wasn't really in control. A follower of God would argue that the chances would be zero. We could fathom a very small number, like one chance in a trillion, but once again I'm going to play it safe and pick one chance in a million.

DIVIDING HIS GARMENTS

It is no exaggeration to say that the Roman soldiers were barbaric in their administration of law and order, particularly when it came to a crucifixion. Their objective was to instill terror and obedience in those watching and to do that with the maximum degree of mutilation and degradation. The Jewish historian Flavius Josephus, in his exhaustive history *The Antiquities of the Jews,* recounted a story about how the Roman soldiers crucified over five hundred people per day, day after day, along the walls of Jerusalem and then left them up for the vultures to consume as they decayed.[46] Only when they ran out of space for more crosses did the crucifixions cease, so intense was their hatred of the Jews. It is against this background that we find an intriguing prophecy fulfilled.

168

A thousand years before Jesus, Israel's King David penned these words suggestive of the crucifixion of Jesus: "*My enemies surround me like a pack of dogs; an evil gang closes in on me. They have pierced my hands and feet...they divide my garments among themselves and throw dice for my clothing*" (Psalm 22:16–18). When we look to the New Testament parallel to this prophecy we see the following account of part of the crucifixion scene, as recorded in the Gospel of Mark: "*Then the soldiers nailed him to the cross. They divided his clothes and threw dice to decide who would get each piece*" (Mark 15:24). This leads us to ask two questions.

Isn't it curious that the very words King David wrote came to pass exactly as he had written them? Furthermore, why would soldiers who held the Jews in such utter contempt want to have the blood-soaked articles of clothing of a Jew they had just murdered? Why indeed! Historians have estimated that during the first century CE, the Romans crucified between fifty thousand and one hundred thousand Jews. What then would be the chances that the clothes of a prisoner would be so coveted—almost like trophies? I should think the odds would be one chance in a hundred thousand, but again, to be on the safe side, I will choose only one chance in a thousand.

NAILED!

Filling in the blanks in the preceding reference to Psalm 22:16, we pick up another messianic prophecy: "*They have pierced my hands and feet.*" This is very interesting. There are no references in either Hebrew historical writings or the annals of the nations that David had dealings with that would suggest that his hands and feet had ever been pierced. This is a familiar literary form used by the Hebrews of David's day to convey the depth of their emotions. In David's day, any other meaning would have been a complete mystery, because crucifixion would not be known until the time of the Persians,[47] some four centuries into David's future.

However, herein also lies a clue to future events. The Hebrews knew their Scriptures contained signs—clues to recognize their Messiah at his coming; they just didn't know when that would be. One of the ways they would identify him would be that his hands and feet would be pierced. Here again, Jesus is the fulfillment of another major prophecy. I should think the odds would be quite small, perhaps one chance in a million, but to give the benefit of a doubt, I will choose one chance in a thousand.

NO BROKEN BONES

King David, not likely realizing the prophetic context of what he was writing, said, "*For the LORD protects the bones of the righteous; not one of them is broken!*" (Psalm 34:20). This is another messianic prophecy that came to pass exactly as written. Before we see its fulfillment we need also to look at God's directions concerning the Sabbath and the handling of dead bodies as the result of execution. We'll start with considerations surrounding the Sabbath.

When God created the earth, he defined that the evening and the morning, in that order, constituted a day: "*God called the light 'day' and the darkness 'night.' And evening passed and morning came, marking the first day*" (Genesis 1:5). To this day, the Jewish day starts at sundown, not at sunrise as it does for most non-Jewish folks. Specific to the Sabbath, God declared, "*Remember to observe the Sabbath day by keeping it holy. You have six days each week for your ordinary work, but the seventh day is a Sabbath day of rest dedicated to the Lord your God. On that day no one in your household may do any work*" (Exodus 20:8–10). Hold on to this while we switch briefly to God's laws about executions.

Concerning the handling of dead bodies, we read, "*If someone has committed a crime worthy of death and is executed and hung on a tree, the body must not remain hanging from the tree overnight. You must bury the body that same day*" (Deuteronomy 21:22–23).

The crucifixion of a Jew on the day before the Sabbath prompted important spiritual necessities; the body had to be properly dealt with so those who had been tending it would not have to work after sundown. Both historic and biblical records clearly confirm that Jesus was crucified the day before the Jewish Sabbath. To ensure he was dead with enough time to get him down from the cross and properly prepare his body for burial before the Sabbath began, the Romans were prepared to break his legs—a routine practice—thereby denying him any ability to raise his body up to breathe, and in so doing making death by asphyxiation imminent. However, when the soldiers saw that he was already dead, they didn't have to break his legs. John was present at the crucifixion and wrote about these events, saying, "*But when they came to Jesus, they saw that he was already dead, so they didn't break his legs*" (John 19:33).

Nor had they broken any other bones, and with that, the prophecy had been fulfilled. I think the probability of them not breaking his legs would be perhaps one chance in a thousand.

AN UNNATURAL DARKNESS

The ministry of the prophet Amos spanned the period from 809 to 784 BCE. It was during this time that he received a revelation from God foretelling a future event linked to the coming of their Messiah: "'*In that day,'* says the Sovereign LORD, *'I will make the sun go down at noon and darken the earth while it is still day'*" (Amos 8:9). Some eight hundred years later, on Passover 33 CE, Jesus Christ of Nazareth was crucified, and on that very day the sun ceased to shine from noon until three in the afternoon. Those of his inner circle recorded the events, each in their own way.

Luke gave his eyewitness account: "*By this time it was about noon, and darkness fell across the whole land until three o'clock. The light from the sun was gone*" (Luke 23:44–45). The darkness described here was not fanciful writing by those who were Jesus' friends; it actually happened and was documented by others outside his followers. Thallus was one of them, although his peers generally derided his scholarly efforts.[48] He was a pagan historian living shortly after the crucifixion. In attempting to explain away the darkness as something completely natural, he contended that the earth darkened because of a solar eclipse.

Julius Africanus, a North African historian familiar with the Jewish traditions, later challenged his flawed treatise and wrote,

> Thallus, in his third book of histories, explains away this darkness as an eclipse of the sun, unreasonably as it seems to me. For the Hebrews celebrate the Passover on the fourteenth day according to the moon, and the passion of our Savior falls on the day

before the Passover; but an eclipse of the sun takes place only when the moon comes under the sun. And it cannot happen at any other time but in the interval between the first day of the new moon and the last of the old, that is, at their junction. How then should an eclipse occur when the moon is almost diametrically opposite the sun?[49]

What about the reality of this event? We can start with three known facts: a solar eclipse can only occur when the moon is directly between the earth and the sun; we know that the crucifixion occurred on a Passover Friday, and Passover, by definition, always took place when the moon was full. But wait, a full moon only occurs when the alignment is in the sun-earth-moon order, not when the order is sun-moon-earth as Thallus proposed. So Thallus had to be wrong.

The prophet Amos prophesied that the darkness would begin at noon; secular historical records confirmed that it did and that it lasted three full hours, just as Luke and others observed. A careful examination of modern solar eclipse data will also lead to the conclusion that full solar eclipses last a maximum of only seven and a half minutes,[50] not three hours as the historic evidence confirms. The conclusion: the eclipse theory must be rejected and the evidence that this strange darkness was a supernatural event accepted. Can we find more evidence?

Phlegon, a Greek historian, also documented this: "With regard to the eclipse in the time of Tiberius Caesar, in whose reign Jesus appears to have been crucified, and the great earthquakes, which then took place, etc."[51]

We can also look at the writings of one Quintus Septimius Florens Tertullianus, an early Christian author[52] more commonly known by the Anglicized name

Tertullian.[53] Concerning the darkening of the sun, he wrote, "In the same hour, too, the light of day was withdrawn, when the sun at the very time was in his meridian blaze.[54] Those who were not aware that this had been predicted about Christ, no doubt thought it an eclipse. You yourselves have the account of the world-portent still in your archives."[55]

And what shall we say then of the probability? It would seem from these additional historic accounts that the darkening of the sun was indeed a true and protracted event. On the strength of these accounts and the fact that such an event is an astronomical impossibility, I would instinctively rank this event with a probability of one chance in a trillion but will settle for one chance in a million. Adjust the odds as you see fit.

BURIAL WITH THE RICH

Here is a most curious prophecy worthy of our consideration. It says that the Messiah would be buried with the rich: "*He was put in a rich man's grave*" (Isaiah 53:9). With respect to Jesus, that's quite a contrast; he had no recorded home or possessions aside from his clothing. He had been flogged like the worst of the criminals the Romans had to deal with, crucified naked between two known thieves, mocked, spat upon and humiliated with insults. But in the end, he was taken down from the cross and buried in a never-before-used tomb belonging to a rich man. In the corresponding New Testament reference Matthew writes,

As evening approached, Joseph, a rich man from Arimathea [a city in Judea] who had become a follower of Jesus, went to Pilate and asked for Jesus' body. And Pilate issued an order to release it to him. Joseph took the body

and wrapped it in a long sheet of clean linen cloth. He placed it in his own new tomb, which had been carved out of the rock. Then he rolled a great stone across the entrance and left (Matthew 27:57–60).

Here we have one more prophecy fulfilled, exactly as stated.

In determining the probability of this event happening, we might ask the question, "How many of Jesus' disciples in Jerusalem at the time of his death were rich enough and willing to donate their burial tomb to someone else?" Maybe one in a hundred?

BAD NEWS FOR THE GUARDS

It's one thing to win a battle that you've fought hard for, and it's quite another to hold on to what you've gained in that battle. And that's exactly what the priests and Pharisees did. Jesus was dead and buried, but there was still one nagging little detail they needed to deal with. While Jesus was alive, he had told his followers that three days after he was crucified he would rise back to life again: *"Then Jesus began to tell them that the Son of Man must suffer many terrible things and be rejected by the elders, the leading priests, and the teachers of religious law. He would be killed, but three days later he would rise from the dead"* (Mark 8:31). That definitely would have been bad news for the Jewish leaders.

Not being dummies, the day after the entombment they arranged for an armed guard to protect the tomb until the three days had passed:

The next day, on the Sabbath, the leading priests and Pharisees went to see Pilate. They told him, "Sir, we remember what that deceiver [Jesus] once said while he was still alive: 'After three days I will rise from the dead.' So we request that you seal the tomb until the third day. This will prevent his disciples from coming and stealing his body and then telling everyone he was raised from the dead! If that happens, we'll be worse off than we were at first." Pilate replied, *"Take guards and secure it the best you can."* So they sealed the tomb and posted guards to protect it (Matthew 27:62–66).

Yet, as we will see in the next section, "He's Back," by the third day the seal was broken, the tomb was empty and the guards were nowhere to be found, with no mention of what happened to the them. Now, we all know that if a soldier falls asleep on guard duty or abandons his post for any reason, he is in very big trouble, even to the likelihood of being executed. The Roman soldiers were not known for their gracious behavior, so I would not imagine that the guards would have left voluntarily. But they did. We might ask the guards what they thought their chance of coming out of this mess alive was. Likely zero, but I'll say one in a thousand.

HE'S BACK!

This last prophecy we'll look at is pivotal to confirming if we've found a creator or not. The fundamental question, I believe, is whether or not Jesus was resurrected. There's a lot to examine and a lot to comment on, so we begin with the Old Testament references.

We start in the book of Psalms where it says, *"For you will not leave my soul among the dead or allow your holy one to rot in the grave"* (Psalm 16:10). King David loved to be in the presence of his Lord God; in the words he penned he likely had no idea how they referred to the coming Messiah. Yet the essence of what he wrote prophesied that the Messiah would die but that God would not allow him to remain in the grave to decay. If the Messiah died and was buried and God didn't allow him to remain in the grave, then where is he? The plot thickens.

Following Jesus' burial, Mary Magdalene—one of Jesus' followers—and Mary the mother of James, one of his disciples, had prepared spices to take to the tomb to anoint the body of Jesus with. We read,

> *But very early on Sunday morning the women went to the tomb, taking the spices they had prepared. They found that the stone had been rolled away from the entrance. So they went in, but they didn't find the body of the Lord Jesus. As they stood there puzzled, two men suddenly appeared to them, clothed in dazzling robes. The women were terrified and bowed with their faces to the ground. Then the men asked, "Why are you looking among the dead for someone who is alive? He isn't here! He is risen from the dead! Remember what he told you back in Galilee, that the Son of Man must be betrayed into the hands of sinful men and be crucified, and that he would rise again on the third day"* (Luke 24:1–7).

First we had the Psalmist's prophecy, and now we have these two men, who are considered by biblical scholars to be angels, standing inside an empty tomb that had been sealed and heavily guarded by Roman soldiers, reminding the women that Jesus had been very specific to them about the events of the previous three days, including the fact that on the third day, the day they were now in, he would rise from the dead. What must have been going through their heads? What skepticism could be going through ours? I dare say we have a murder mystery on our hands that demands proof. Stay open. Read on.

The scene now switches to two of Jesus' disciples walking on the road to Emmaus, a town not far from Jerusalem. As they journeyed along, they were discussing all the events of the past few days that had culminated in Jesus' crucifixion and the discovery three days later that the sealed tomb he had been buried in had been inexplicably opened right under the noses of the Roman soldiers posted to secure it. And the body—gone! It wasn't long before Jesus joined up with them, as we read in Luke's account:

That same day two of Jesus' followers were walking to the village of Emmaus, seven miles from Jerusalem. As they walked along they were talking about everything that had happened. As they talked and discussed these things, Jesus himself suddenly came and began walking with them. But God kept them from recognizing him. He asked them, "What are you discussing so intently as you walk along?" They stopped short, sadness written across their faces. Then one of them, Cleopas, replied, "You must be the only person in Jerusalem who hasn't heard about all the things that have happened there the last few days." "What things?" Jesus asked. "The things that happened to Jesus, the man from Nazareth," they said. "He was a prophet who did powerful miracles, and he was a mighty teacher in the eyes of God and all the people. But our leading priests and other religious leaders handed him over to be condemned to death, and they crucified him. We had hoped he was the Messiah who had come to rescue Israel. This all happened three days ago. Then some women from our group of his followers were at his tomb early this morning, and they

178

came back with an amazing report. They said his body was missing, and they had seen angels who told them Jesus is alive! Some of our men ran out to see, and sure enough, his body was gone, just as the women had said." Then Jesus said to them, "You foolish people! You find it so hard to believe all that the prophets wrote in the Scriptures. Wasn't it clearly predicted that the Messiah would have to suffer all these things before entering his glory?" Then Jesus took them through the writings of Moses and all the prophets, explaining from all the Scriptures the things concerning himself. By this time they were nearing Emmaus and the end of their journey. Jesus acted as if he were going on, but they begged him, "Stay the night with us, since it is getting late." So he went home with them. As they sat down to eat, he took the bread and blessed it. Then he broke it and gave it to them. Suddenly, their eyes were opened, and they recognized him. And at that moment he disappeared! They said to each other, "Didn't our hearts burn within us as he talked with us on the road and explained the Scriptures to us?" And within the hour they were on their way back to Jerusalem. There they found the eleven disciples and the others who had gathered with them, who said, "The Lord has really risen! He appeared to Peter." Then the two from Emmaus told their story of how Jesus had appeared to them as they were walking along the road, and how they had recognized him as he was breaking the bread (Luke 24:13–35, emphasis added).

This adds two more eyewitnesses who saw and talked with the risen Jesus. Whom else can we find to testify to the authenticity of the resurrection claim? Luke had more to say in that passage, so we continue to look there.

As I was reflecting on this last paragraph, it suddenly

dawned on me why Jesus *"took them through the writings of Moses and all the prophets, explaining from all the Scriptures the things concerning himself."* It was because Moses and the prophets had recorded so many messianic prophecies that it would be impossible not to recognize the Messiah when he came. And here we have Jesus revealing to his friends that the evidence of who he is lines up with those prophecies, without exception. In a modern-day court of law, this much supporting evidence would be more than ample to arrive at a definitive verdict.

As the two travelers were reporting their encounter to the disciples back in Jerusalem, Jesus appeared to them, as Luke goes on to say:

> *And just as they were telling about it, Jesus himself was suddenly standing there among them. "Peace be with you," he said. But the whole group was startled and frightened, thinking they were seeing a ghost! "Why are you frightened?" he asked. "Why are your hearts filled with doubt? Look at my hands. Look at my feet. You can see that it's really me. Touch me and make sure that I am not a ghost, because ghosts don't have bodies, as you see that I do." As he spoke, he showed them his hands and his feet. Still they stood there in disbelief, filled with joy and wonder. Then he asked them, "Do you have anything here to eat?" They gave him a piece of broiled fish, and he ate it as they watched. Then he said, "When I was with you before, I told you that everything written about me in the law of Moses and the prophets and in the Psalms must be fulfilled"* (Luke 24:36–44; see also John 20:19–20).

So here we have at least eight additional witnesses.

Next, we see Jesus appearing to Thomas, the only one of his disciples not there when he appeared to the others in Jerusalem. John records that event in his Gospel:

> *One of the twelve disciples, Thomas (nicknamed the Twin), was not with the others when Jesus came. They told him, "We have seen the Lord!" But he replied, "I won't believe it unless I see the nail wounds in his hands, put my fingers into them, and place my hand into the wound in his side." Eight days later the disciples were together again, and this time Thomas was with them. The doors were locked; but suddenly, as before, Jesus was standing among them. "Peace be with you," he said. Then he said to Thomas, "Put your finger here, and look at my hands. Put your hand into the wound in my side. Don't be faithless any longer. Believe!" "My Lord and my God!" Thomas exclaimed* (John 20:24–28, emphasis added; see also Mark 16:14).

Many of us can easily identify with Thomas; we hunger for proof, especially when something is so evident to others, but some conclusive detail they are privy to is not yet available or evident to us. Poor Thomas; I can only guess how the others must have persisted in trying to convince him that Jesus really was alive. What must his thoughts have been? Talk about inner turmoil! Tally one more eyewitness in Thomas.

One day soon after his resurrection, Jesus manifested himself to seven of his disciples as they were fishing:

> *Later, Jesus appeared again to the disciples beside the Sea of Galilee. This is how it happened. Several of the disciples were there—Simon Peter, Thomas (nicknamed the Twin), Nathanael from Cana in Galilee, the sons of Zebedee, and*

two other disciples. Simon Peter said, "I'm going fishing."
"We'll come, too," they all said. So they went out in the boat,
but they caught nothing all night. At dawn Jesus was
standing on the beach, but the disciples couldn't see who he
was. He called out, "Fellows, have you caught any fish?"
"No," they replied. Then he said, "Throw out your net on the
right-hand side of the boat, and you'll get some!" So they did,
and they couldn't haul in the net because there were so many
fish in it. Then the disciple Jesus loved said to Peter, "It's the
Lord!" (John 21:1–7, emphasis added).

The events so far have recorded that Jesus appeared to
small numbers here and there, but that was soon to
change, dramatically. In his first letter to the converts to
Christianity in Corinth, Greece, the apostle Paul gives us
insight to five hundred more who saw Jesus alive following
his resurrection:

Let me now remind you, dear brothers and sisters, of the
Good News I preached to you before. You welcomed it then,
and you still stand firm in it. It is this Good News that saves
you if you continue to believe the message I told you—unless,
of course, you believed something that was never true in the
first place. I passed on to you what was most important and
what had also been passed on to me. Christ died for our sins,
just as the Scriptures said. He was buried, and he was raised
from the dead on the third day, just as the Scriptures said.
He was seen by Peter and then by the Twelve. After that, he
was seen by more than 500 of his followers at one time, most
of whom are still alive, though some have died. Then he was
seen by James and later by all the apostles. Last of all, as
though I had been born at the wrong time, I also saw him
(1 Corinthians 15:1–8).

Conclusion? There were hundreds who could testify to a living, breathing Jesus after his crucifixion. If the writers who chronicled these facts failed to present them accurately, wouldn't their integrity have been challenged in very short order?

WHERE NOW?

It would then seem that the appearances of Jesus were well corroborated by a large number of witnesses; but where is Jesus now? For the answer to that, we go to the Gospel of Luke: "*Then Jesus led them to Bethany,*[56] *and lifting his hands to heaven, he blessed them. While he was blessing them, he left them and was taken up to heaven*" (Luke 24:50-51).

A similar account is given in Mark's Gospel: "*When the Lord Jesus had finished talking with them, he was taken up into heaven and sat down in the place of honor at God's right hand*" (Mark 16:19).

If we have trouble believing Jesus is no longer on earth but in heaven, is there any evidence to support this idea besides the written accounts just presented? Has even one account been substantiated that Jesus has been seen alive and on planet Earth after his (so-called) ascension? No! Not one has ever been documented, so it would appear not.

This, perhaps the most publicized event in human history, has no evidence to support that it did not happen. Think how fast the discovery of Jesus' body would have delighted his opponents; it would have given them all they needed to expose the entire affair as a hoax. But that discovery has never happened. I know this is heavy stuff that not all of us will be able to accept.

It's time to summarize what we have and see what we can conclude from it:

183

- There were multiple eyewitnesses to the crucifixion, the burial and the empty tomb.
- There were hundreds who saw Jesus alive afterwards.
- There were multiple firsthand accounts to his ascension to heaven.
- His body has never been found.
- No one has ever disproved any of these events.
- No one has ever successfully challenged the writings of the New Testament writers who knew Jesus personally as being fraudulent in any way.

To bring these considerations to a close, I present two quotes for your consideration.

The first is from J. N. D. Anderson, a respected scholar and historian:

The most drastic way of dismissing the evidence would be to say that these stories were mere fabrications, that they were pure lies. But, so far as I know, not a single critic today would take such an attitude. In fact, it would really be an impossible position. Think of the number of witnesses, over five hundred. Think of the character of the witnesses, men and women who gave the world the highest ethical teaching it has ever known, and who, even on the testimony of their enemies, lived it out in their lives. Think of the psychological absurdity of picturing a little band of defeated cowards cowering in an upper room one day and a few days later transformed into a company of witnesses that no persecution could silence, and then attempting to attribute this dramatic change to nothing more convincing than a miserable fabrication they were trying to foist upon the world. That simply wouldn't make sense.[57]

The second is from Chuck Colson, who was deeply implicated in the Watergate scandal during U.S. President Richard Nixon's term in office.[58] Following his incarceration, Colson became an outspoken Christian advocate. In his following quote, he points out just how difficult it would have been to successfully sustain a lie for any protracted period of time:

> I know the resurrection is a fact, and Watergate proved it to me. How? Because twelve men testified they had seen Jesus raised from the dead, and then they proclaimed that truth for forty years, never once denying it. Every one was beaten, tortured, stoned and put in prison. They would not have endured that if it weren't true. Watergate embroiled twelve of the most powerful men in the world, and they couldn't keep a lie for three weeks. You're telling me twelve apostles could keep a lie for forty years? Absolutely impossible![59]

So now we must be in agreement on a probability that these events took place. What are the chances that this Jesus actually was not raised from the dead as the prophets of Israel foretold? The evidence seems so conclusive that I lean toward a number that is so minutely small as to be zero. For the final time, I defer to a more finite number, one chance in a million. As always, pick your own probability if you prefer.

And just where has this journey led us?

Last But Not Least

In the earlier part of this book, we considered over one hundred questions challenging evolution and got no substantiating evidence to support it. And now we look at the odds supporting the premise of Jesus being the creator and the Messiah spoken of by ancient Israel's prophets.

We've just finished looking at only eighteen of the forty-eight[60] major Old Testament prophecies that pointed to a Jewish Messiah. If we put the odds of just these few together, we end up with two sets of probabilities, depending on the ancestral lineage of Jesus through Mary or Joseph. To get these final probabilities, we must multiply the individual probabilities together; e.g., to get the probability of Jesus being the Messiah through Joseph's line, we multiply one chance in two followed by twenty-two zeros (for Jesus being born through Joseph's lineage), then multiply that by one chance in twelve (Jesus' chance of coming from the tribe of Judah), then multiply again by one chance in fifty (his chances of being born in

Bethlehem), etc., for all the probabilities we explored in the previous section, "For Thinkers Only."

When we are finished multiplying all those probabilities of Jesus precisely fulfilling these eighteen prophecies through Joseph's ancestral line, we get:

1 chance in 700,000,000,000,000,000,000,000,000, 000,000,000,000,000,000,000,000,000,000,000, 000,000,000,000,000.

If we go through the ancestral line of Jesus' mother, Mary, and apply the probabilities as we did for Joseph, then for the probability that Jesus precisely filled the same eighteen prophecies we get:

1 chance in 6,000,000,000,000,000,000,000,000,000, 000,000,000,000,000,000,000,000,000,000,000, 000,000,000,000,000,000,000.

These are huge numbers that support that Jesus is exactly whom the Old Testament prophets foretold Israel's Messiah and the Creator[61] of the heavens and the earth would be.

Here's the bottom line: if you stayed with me this far, you can see the numbers for yourself. Maybe you didn't use the same probabilities I did, and that's OK. Whatever you used, tally them and see what you get.

Don't forget, with both our study of the human body being the result of a creative process and our study of Jesus being the Creator of "*the heavens and the earth*" (Genesis 1:1), the evidence shown was only a tiny fraction of what could have been presented. What do you think the odds would look like if all the facts and the forty-eight major prophecies had been fully explored? I'm not a betting man, but I can recognize what is tantamount to a sure thing when I see it. Such incomprehensible numbers necessitate one more consideration.

The Bible's New Testament writers, as well as secular authors of the first century CE, have certainly documented indisputable evidence to support Jesus as the fulfillment of the messianic prophecies. Some of you will be saying to yourself, "Wow! That's for sure; I want to know more," and others will be saying, "So what?" Both are completely viable comments. All along, I've been presenting possibilities with which you can agree or disagree; I want the final decisions—those based on the collective probabilities—to be yours, not mine.

OPTIONS

At this point there likely could be three courses that you, the reader, might elect to take from here. The first would be to say, "Well, that was interesting," and put the book on the shelf. The second would compel you to want to know more about the Bible and Jesus. And the third would challenge your personal allegiance to another belief system or being—religion if you will—the entity or philosophy in which, or in whom, you base your belief in an afterlife. Let's briefly consider each of these options.

THE "ON THE SHELF" OPTION

Even if you dismiss the evidence that has been offered, there is a "however" you must be aware of. You've read some pretty challenging stuff, and because of that, eventually you will need to come to a resolve with it—a position you can live with. I believe a line has been drawn in the sand for you, a line you must sooner or later come to grips with. You see, these facts will eventually demand a resolution.

Can you dispute the evidence against evolution and embrace it instead of creation? If you do, can you logically also dismiss the massive evidence in support of a Creator? Sooner or later, having been presented with these facts, we must all answer two questions: "How am I going to handle the potential reality that there really is a God?" and "What am I going to do about Jesus?"

Many readers may want to do a "full stop" at this point. They don't want to believe in God, because they don't want to be accountable to him; they want to do what they want to do, living by their own rules. They want their own brand of religion so they are accountable solely to themselves. If this is you, I urge you to read on; there's so much more for you.

THE "KNOW MORE" OPTION

Allow me a slight digression in order to make a point. I have a degree in electrical engineering, but, you know, when it comes to hooking up stereo systems, HDTVs, etc., I am a typical guy—I ignore the instructions and do my own thing, usually to my chagrin. For example, several years ago we bought the whole nine yards: TV, surround sound stereo, CD/DVD player, tape deck, radio, link to our aging record player, video game console, etc. Each came with its own thick instruction manual, so my first step was to pile the manuals on the couch; after all, I do have a university degree in all this stuff! My next step was to start wiring everything together, as only a know-it-all could do.

I don't need to tell you what happened; doubtless you've been reading between the lines. I'd press a button to start the TV, and the tape deck would fire up; I'd press another button to turn on the DVD deck, and the stereo would come to life. Not a great testimony to four long, expensive years in university! By ignoring the manufacturer's handbooks, I messed up, not only on installing the individual components but also on the installation of the system as a whole. Oh, some things worked, but more didn't than did. Way to go, Bill!

And the point is...? If I had read the manufacturer's instruction manuals I would have had a successful installation and years of trouble-free operation.

And the point of this anecdote is...?

The point is this: if we accept that Jesus is exactly who the prophets described him to be and that he fully fits the description of the one who "*created the heavens and the earth*," then would we not be well advised to consider what else we could find in the Creator's handbook, the Bible? In John's Gospel, Jesus himself encouraged his audience when he said, "*You are truly my disciples if you remain faithful to my teachings. And you will know the truth, and the truth will set you free*" (John 8:31–32).

When I first stopped to look at the Bible as a whole, rather than just a source of prophecy-here and fulfillment-there events, it seemed like a daunting volume. Trying to make sense out of it loomed like a "mission impossible" exercise; however, with a little guidance from folks who knew it rather intimately, I was able to make sense out of it. Let me share the experience.

To begin with, there are sixty-six books in it: thirty-nine in the Old Testament and twenty-seven in the New Testament. Each is divided into chapters, then subdivided into verses. In my voyage of discovery, a number of Bible verses struck a chord. Let me try to put them in some kind of order.

A LOVE STORY

The first verse that really got to me said, "*God is love*" (1 John 4:8). I think all of us can use as much love as we can get; I know I can. Do you remember an earlier verse that said that Jesus created "*all things*" (John 1:1–3 NASB)? That would include you and me. Add these together and the immediate conclusion would be that Jesus loves His creation, too.[62]

There were other "love" verses I found where God speaks directly of His love nature:

> "*You are precious to me. You are honored, and I love you*" (Isaiah 43:4).

> "*I have called you by name; you are mine*" (Isaiah 43:1).

> "*I have loved you, my people, with an everlasting love*" (Jeremiah 31:3).

> *The LORD is merciful and compassionate, slow to get angry and filled with unfailing love* (Psalm 145:8).

> *O Lord, You are so good, so ready to forgive, so full of unfailing love for all who ask for your help…your love for me is very great* (Psalm 86:5,13).

The more I read the Bible, the more it strikes me as a love letter from God to the human race.

DECEIVED

Why is it that so many of us never knew of this Creator God, this God that says He loves us? Why has He been so hidden from us? The Bible speaks to that issue as well. In the very first chapter of the Bible, we read the description of God's creation of the heavens and the earth and His opinion of it: it was "*good.*" It also describes His creation of the first humans, Adam and Eve; they were *"very good."* Chapters 1 and 2 then describe the domain that God created for them as a dwelling place, the Garden of Eden. God had told Adam and Eve they could have free access to every fruit-bearing tree in the Garden, save one.

The Bible also talks about God's most glorious created angel, Lucifer (more commonly known to us today as Satan, the devil), his ambitions to usurp God's power and his throne, and God's judgment upon him:

"How you are fallen from heaven, O shining star, son of the morning!...For you said to yourself, 'I will ascend to heaven and set my throne above God's stars. I will preside on the mountain of the gods far away in the north. I will climb to the highest heavens and be like the Most High.' Instead, you will be brought down to the place of the dead, down to its lowest depths" (Isaiah 14:12–15).

Satan's tactic was to destroy everything God had created, especially humankind. Knowing the constraint God had placed on what Adam and Eve could eat, Satan took no time in convincing them that they, too, could be like God, as soon as they ate the forbidden fruit. His smooth-talking lies quickly seduced Adam and Eve into disobeying God's directive. The Bible called this disobedience sin, and there were consequences Adam and Eve would pay for it, foremost being their expulsion from the Garden of Eden and a significant change in the intimate relationship they had enjoyed with God. God had created mankind to be immortal; it was sin entering the human race through Satan's treachery that brought death as its companion.

In the Bible Satan is called the deceiver, a name that aptly describes his dealings with the likes of you and me:

"*This great dragon—the ancient serpent called the devil, or Satan, the one deceiving the whole world*" (Revelation 12:9). He wants us to have absolutely no understanding of the conditions God has placed on us for sharing a glorious joy-filled eternity with Him. Small wonder that we are so ignorant of the great promises the Bible holds for us if we will only believe; Satan is in an all-out battle to depose God and keep us from eternity with Jesus, and he is using us as the instruments to reach his ends.

Consider this. Go to a movie, watch TV, listen to people arguing, and what are you likely to hear? When people curse or voice their vehement disdain for something or someone, why is it always "God" or "Jesus" or "Christ" that are the universally preferred swear words they use? Why are Allah or Buddha or Confucius or the Dalai Lama or Shiva or others never mentioned? Curious, isn't it?

Attending university in a very cosmopolitan city meant I had friends from many different nationalities and varied religious backgrounds; it also meant I knew how to swear in most of their languages. Regardless of the object of their spiritual focus, when they wanted to swear, their expressions would inevitably include God, Jesus or Christ. Doesn't that sound like something God's archenemy would inspire in those who are not yet Christians? I understand why people use these three specific names in their chosen profanities: they are Satan's pawns to voice his utter contempt toward the God who loves His creation and wants a personal relationship with them. It really makes you stop and think, doesn't it?

WAR

When I think back over my life and the questions I've asked, I think of the countless friends and acquaintances I've known who have asked the same probing questions. What's the purpose of life? Why am I here? There must be more to life; what is it? How do I find it? Why do I feel a great emptiness within me, like there's something missing? These questions, and others like them, plagued me for years.

Now I see that Satan's goal is to anaesthetize the human soul to any understanding of God, striving to make Him virtually irrelevant. It's a war between Satan and God that God will most certainly win; yet Satan is determined to unleash his subtle treacheries any way he can to maximize the number of souls he will prevent from seeking God's love— souls that he will hold captive for eternity. How, then, can we come out of this battle on God's winning side?

THE RESCUE PLAN

God wants us to be free of sin and to be with Him, not just in our human lifetime but forever, and Jesus made no bones about it: "*There is more than enough room in my Father's home. If this were not so, would I have told you that I am going to prepare a place for you? When everything is ready, I will come and get you, so that you will always be with me where I am*" (John 14:2–3). Jesus' birth, death and resurrection were God's rescue plan to cancel the curse of sin.

This plan has been recorded in John's Gospel: "*For God loved the world so much that he gave his one and only Son [Jesus], so that everyone who believes in him will not perish but have eternal life*" (John 3:16).

Paul put it this way: "*For the wages of sin is death, but the free gift of God is eternal life through Christ Jesus our Lord*"

(Romans 6:23). This verse is saying that Jesus is offering each of us the gift of eternal life, but if we do not reach out and take it, it never becomes ours; it's just an unaccepted gift, and we remain in a state of sin—spiritual death and eternal separation from God. There will be more on this farther on.

WHAT SIN?

Do we think for a minute that we are not sinful? Satan works relentlessly to have us believe that we are not. Everything he does is a direct effort to discredit God and His Word, so we need to be very careful here. For example, there was a time when I certainly didn't think I was sinful, but after searching out what the Bible has to say about it, ouch!

When I began to wonder if the whole Bible was true, not just the prophecies about Jesus, here are just a few of the things I discovered that God says are sins:

having other gods before Him
worshiping idols of any kind (whether that be your car, boat, job, boyfriend, girlfriend, skills, etc.)
misusing His name
misusing His day of rest—the Sabbath
dishonoring parents
murdering
committing adultery
stealing
giving false testimony against anyone
coveting another person's wife or husband or anything he or she possesses
fornicating
men behaving as a woman (effeminate)

practicing homosexuality or lesbianism
being drunk
reviling
swindling

When I first saw this list, it was like an arrow through my heart—convicting! For years, I had trusted in me and my bank account more than in God, used His name when cursing or swearing, worked on the Sabbath, dishonored my parents, lusted after attractive women, stolen things, and on and on. Not very pretty, I'm afraid.

I happened to be watching a TV documentary a few weeks ago in which the interviewer was posing questions to strangers in a town square, somewhere in Europe. His opening questions to each individual were "Have you ever sinned? Do you believe you are a sinner?" Their immediate answer was, always, "No."

Next question: "Have you ever stolen anything?" Inevitably the answer was a sheepish affirmative. "Have you ever sworn using the names God, Jesus or Christ?" A bit of a blush, and "Yes." And on it went.

Six or seven questions later the interviewer asked if they still thought that they had never sinned. There was an embarrassing pause from each, and they admitted that they had sinned, as the Bible defines it. The Bible confirmed my understanding where it says, "*For everyone has sinned; we all fall short of God's glorious standard*" (Romans 3:23).

GRACE PROMISED

How then can we, with sin, come close to God? The answer to that can be found in the reason why the Messiah Jesus came to earth. Paul, in his letter to the church at Ephesus,[63] put it all together succinctly:

Once you were dead because of your disobedience and your many sins. You used to live in sin, just like the rest of the world, obeying the devil—the commander of the powers in the unseen world. He is the spirit at work in the hearts of those who refuse to obey God. All of us used to live that way, following the passionate desires and inclinations of our sinful nature. By our very nature we were subject to God's anger, just like everyone else. But God is so rich in mercy, and he loved us so much, that even though we were dead because of our sins, he gave us life when he raised Christ from the dead. (It is only by God's grace that you have been saved!) For he raised us from the dead along with Christ and seated us with him in the heavenly realms because we are united with Christ Jesus. So God can point to us in all future ages as examples of the incredible wealth of his grace and kindness toward us, as shown in all he has done for us who are united with Christ Jesus. God saved you by his grace when you believed. And you can't take credit for this; it is a gift from God. Salvation is not a reward for the good things we have done, so none of us can boast about it (Ephesians 2:1–9).

The last two sentences in this Scripture make a most liberating statement. Good works cannot save us from our sins; only God's unmerited grace when you believe in Jesus.

Consider also what we find in John's Gospel: "*But to all who believed him and accepted him, he gave the right to become children of God*" (John 1:12).

Ah, but there must be other ways to God, one might argue. That reasoning takes us back again to the infallibility of the Bible; if we are convinced that the evidence presented proves that Jesus is the Messiah and the Creator of all, then His words must be taken at face value when He says, "*I am the way, the truth, and the life. No one can come to the Father [God] except through me*" (John 14:6). And the "*through me*" happens when we receive Him into our lives as Savior and *Lord.*

This receiving Him is not a head thing; nor is it emotional or simply saying the words, like some kind of magical formula. No, Ephesians 2:5 says we are saved by His grace. If we receive Jesus, we are asking Him to pardon our sins and take over ownership of our lives from here on, directing us in His way. If we do, Paul's letter to the church in Corinth[64] tells us what awaits us: "*So all of us who have had that veil removed can see and reflect the glory of the Lord. And the Lord—who is the Spirit—makes us more and more like him as we are changed into his glorious image*" (2 Corinthians 3:18). In other words, once we have accepted Jesus, He begins to transform our lives to be like His.

Here's another promise God made to His people, "'*I know the plans I have for you,' says the LORD. 'They are plans for good and not for disaster, to give you a future and a hope*'" (Jeremiah 29:11). That sounds like a "good God," doesn't it?

IT'S NOT ALL ABOUT ME

Why would anyone not want to invite Jesus into his or her life? Fair question. I know what my reasoning was when I was first confronted with all this evidence of a Creator. I simply didn't want to believe in God, regardless of the evidence before me, because if I did I would have to be accountable to Him, and that meant giving up a lot of sinful things in my life that I quite enjoyed, thank you very much. I wanted to do what I wanted to do and do it my way, so I invented my own religion, if you will, complete with what I deemed to be acceptable behavior. In other words,

I considered myself to be a "good" person, regardless of the sins in my life that I knew were in direct opposition to God. That way, I reasoned that if I never acknowledged Him for who He was, I wouldn't have to be accountable to Him. Bad logic!

If that's where you are, let me share another section of New Testament Scripture with you that I discovered. I said, "Ouch!" when I saw this one, too:

But God shows his anger from heaven against all sinful, wicked people who suppress the truth by their wickedness. They know the truth about God because he has made it obvious to them. For ever since the world was created, people have seen the earth and sky. Through everything God made, they can clearly see his invisible qualities—his eternal power and divine nature. So they have no excuse for not knowing God. Yes, they knew God, but they wouldn't worship him as God or even give him thanks. And they began to think up foolish ideas of what God was like. As a result, their minds became dark and confused. Claiming to be wise, they instead became utter fools. And instead of worshiping the glorious, ever-living God, they worshiped idols made to look like mere people and birds and animals and reptiles. So God abandoned them to do whatever shameful things their hearts desired. As a result, they did vile and degrading things with each other's bodies. They traded the truth about God for a lie. So they worshiped and served the things God created instead of the Creator himself, who is worthy of eternal praise! Amen. That is why God abandoned them to their shameful desires. Even the women turned against the natural way to have sex and instead indulged in sex with each other. And the men, instead of having normal sexual relations with women, burned with lust for each other. Men did shameful things with other men, and as a result of this sin, they suffered within themselves the penalty they deserved. Since they thought it foolish to acknowledge God, he abandoned them to their foolish thinking and let them do things that should never be done. Their lives became full of every kind of wickedness, sin, greed, hate, envy, murder, quarreling, deception, malicious behavior, and gossip. They are backstabbers, haters of God, insolent, proud, and boastful. They invent new ways of sinning, and they disobey their parents. They refuse to understand, break their promises, are heartless, and have no mercy (Romans 1:18–31).

Consider a statement Jesus made to folks just like you and me in the last book in the Bible: "*Look! I stand at the door and knock. If you hear my voice and open the door, I will come in, and we will share a meal together as friends*" (Revelation 3:20). Notice that He is knocking on your heart, so to speak, and when you hear Him, it's your choice whether to respond to Him or not. He never forces a decision; He waits for you to come to Him. Talk to Him; He is eagerly waiting to share His very presence with you.

TIME TO SAY "YES"

If this leaves you wanting to know Jesus in a personal way, not just knowing about Him, but needing some additional encouragement, let me suggest the following as an appropriate prayer:

"Jesus, I believe You are the Son of God and that You want to welcome me into Your family. I believe You died on the cross to bear my sins so I would be able to have eternal fellowship with You, and that three days later You rose back to life, conquering death. Please forgive my sins. I trust You to lead me for the rest of my life and to change me to be like You."

If you've prayed this from your heart, He will do exactly as you have asked.

If you don't already have a Bible, I encourage you to get one, either a New Living Translation or a New American Standard Bible, Revised. Now, go find a church that strives to be like Jesus in everything they do; introduce yourself to the church leader and tell him you are a new Christian and want to meet mature Christians who can help mentor you in the ways of God.

THE "OTHER BELIEF SYSTEM" OPTION

By now you may be weighing the various conclusions drawn here. You may also be one of those who say, "Forget it! I don't need this." Wait! Before you turf this, can I give you some challenging thoughts you might want to consider?

What you believe is critically important. Centuries ago, people believed the sun was the center of the universe, but that belief was eventually proven to be wrong. They also believed the earth to be flat—again, disproved. Millions of people have grown up believing there is no God, regardless of the evidence. They hold fast to the beliefs they have been taught or inherited from their families or friends. And what if those beliefs are determining your eternity after death?

Many flatly refuse to believe otherwise than the faith of their peers or fathers. That's what they've chosen to believe, and there's no option for assessing its validity. Some consider religion, especially Christianity, to be all about rules and controls; Jesus promotes no such thing. He alone promises you a personal and loving relationship with Him, both in this life and for eternity. Acknowledging Him as the *Lord* of your life brings huge benefits, just a few of which are listed at the beginning of the next section.

Only Jesus has the authority to forgive your sins—past, present and future—and to present you blameless to His Father, God, at the end of your earthly life, or, if you prefer, at the beginning of eternity. Of all the other deities mankind worships, only Jesus rose from the dead and promised His followers that they would too.

Who or what is the object of your worship? Is he or she or it alive or dead and buried? Why would anyone want to base their eternity on someone or something that couldn't possibly support a promise of eternal life when they, themselves, didn't have the power and authority to cheat death? Let me pose one possible answer to this condition.

Have you ever intensely disliked or mistrusted someone you barely knew? I have. What ever drives us to have such strong negative emotions about someone we hardly know? I remember one man vividly; I harbored a huge, albeit irrational, anger in my heart toward him. There was something about him I couldn't trust, and I was only too willing to voice my vehement thoughts about him to anyone who would listen. That was, until the day I met him in a personal way and discovered, much to my shame, how genuine and good he was. Then my heart ached for the lies I had fed my spirit with about him. I had immediate admiration for this man whom just days before I had intensely disliked. I had been terribly wrong.

Maybe you feel the same way about Jesus, yet without reason. Could you rethink where those emotions come from and give Him another chance?

If your choice is still "No, thanks," may I leave you with two encouragements?

FIRST ENCOURAGEMENT

The Bible says the fruit of a godly spirit is "*love, joy, peace, patience, kindness, goodness, faithfulness, gentleness, and self-control*" (Galatians 5:22–23). If this is not how your life is but how you would like it to be, then why not pick up a New Living Translation of the Bible and start to read it? The Bible used in Christian churches has been unchanged since its contents were originally penned under God's guidance.[65] It is a fountainhead of information that will introduce you to the One who not only created you but also wants to have a loving relationship with you.

Read the New Testament first, starting with the Gospel of John, then Psalms, Proverbs and the Old Testament, in that order. You'll be on your way to the greatest adventure you could ever dream of. This Creator is not a hidden, impersonal being, as many consider Him to be. All you need to do is ask Him to lead you; you could say something like "Jesus, please reveal Yourself to me so I can know without a doubt that You are exactly who You claim to be." When you think about it, you've got nothing to lose and everything to gain.

SECOND ENCOURAGEMENT

Even if you dismiss the evidence offered in this book, the facts will never leave you. Why? Because the Scriptures referenced are the direct Word of God and His word is never wasted. Sooner or later, because we've been presented with the facts, we must all answer the question "What am I going to do about Jesus?" I encourage you to address that issue sooner rather than later.

Seek the Lord while you can find him. Call on him now while he is near. Let the wicked change their ways and banish the very thought of doing wrong.

Let them turn to the Lord that he may have mercy on them. Yes, turn to our God, for he will forgive generously.

"My thoughts are nothing like your thoughts," says the Lord. "And my ways are far beyond anything you could imagine.

For just as the heavens are higher than the earth, so my ways are higher than your ways and my thoughts higher than your thoughts.

"The rain and snow come down from the heavens and stay on the ground to water the earth.

They cause the grain to grow, producing seed for the farmer and bread for the hungry.

It is the same with my word. I send it out, and it always produces fruit.

It will accomplish all I want it to, and it will prosper every-where I send it. (Isaiah 55:6-11)

THE "FOR THOSE OF OTHER PERSUASIONS" OPTION

The only sources of evidence I could find that provided all the characteristics I needed to begin this investigation were the writings and historic records of the ancient Hebrews and other peoples who knew and interacted with them. They have provided us with a huge repository of evidence. If you are of a different persuasion than Christianity, then I encourage you to dig into whatever wells of evidence you have at your disposal to see if the one you follow and bear allegiance to has more reliable credentials than what we have found.

Jesus made many profound assurances about life in both the here-and-now and in the spirit after death. Wonderful promises! What I wrote a few paragraphs back deserves to be reiterated; the Bible says the fruit of a godly spirit is "*love, joy, peace, patience, kindness, goodness, faithfulness, gentleness, and self-control*" (Galatians 5:22–23). What does the one you worship have to offer? Can their promises be substantiated? Are you worshipping a lifeless idol or an ideal? Is this life all there is, or is there more, much more? What does your spirit speak to you about these issues? Are you willing to gamble your eternity on any "deity" that offers less than what we have seen here?

Let me close with this little window into my past. I was not the nicest guy Jesus ever had to forgive. I broke most of the commandments regularly. One day I met a pretty nursing student at a university dance and wanted to make her my life's ambition. Big problem, though: she got very little time away from school except every Sunday, which she spent with her parents and going to church. Church? Yah, church. And the only way I could get time with her was if I went to church with her.

I wrestled with it but quickly caved in. After all, I didn't have to listen to what was going on or be a part of it; I just had to be with her. I did it. One of the first things I noticed about her church leader was that he had been a civil engineer for several years before entering the ministry. I immediately set my goal as bringing this poor chap out of the ministry and back into the life of a professional engineer.

Weeks came and went. Sunday morning, Sunday night...Sunday morning, Sunday night. The people in this church all seemed to have this incredible joy that stemmed from their personal relationship with Jesus—all except me, that is. I was still securely entrenched in breaking multiple commandments most weeks. Little did I realize how much of an effect these Sunday sessions were having on me.

One night I decided it was time to reel this leader in and get him free from church and back to being an engineer. I asked him if I could talk with him after the service, and he said OK. When we got together, I had my plan all mapped out. "I have a question for you," I said.

"OK," he replied, "but first can I ask you a question?"

You're not playing fair, I thought; after all, this was my meeting. Anyway, to be gracious I said, "OK"; what importance could his question possibly have compared to what I was about to unleash?

With a measure of love I'd never experienced from anyone before, he simply asked, "Bill, what are you going to do about Jesus?"

That broke the dam; in an instant his question had touched me in the deepest part of my being. I knew exactly what I was about to do, but I fought against it with everything I had in me. Then, realizing that this was both what I needed and what I desperately wanted, I asked Jesus to take over in my life.

I have no idea what happened next except that I sobbed my guts out for what seemed like a very long time. How embarrassing! Guys don't cry! When it was all over, I felt clean, wonderfully clean, inside and out, miraculously free of all the guilt and sins I'd been carrying around. I headed home that night two feet off the ground, and it's been the same ever since.

Do I still sin? Yes, but not at all like I used to; certainly not intentionally. I'm not perfect; that'll come when I meet Jesus face to face. Right now, in this life, I'm forgiven but still a work in process. Most of the old ways drained away from me in the next few years, some almost immediately and others more gradually. Jesus has His ways to rework you. Did it hurt? Never! Would I have done it differently? Yes, sooner, much sooner. You see, now I literally have a two-way conversation with the Creator every day of my life. That is priceless, absolutely priceless.

I had set out to turn this leader around, and he—or should I say, Jesus working in him—turned me around. Let Jesus turn you around too. If you truly want His help, you could say something like "Jesus, please make Yourself real to me, forgive my sins, and take over my life." He will! And you will never regret it.

Never!

P.S. I didn't get the girl, but I did get Jesus.

Epilogue

There came a day in the 21st century when learned and wise scientists had successfully created a human being without the help of a women's egg or a male's sperm. Elated with their achievement, they felt they had overcome the last barrier to permanently declaring that Jesus was no longer of any importance and should be told so.

To make the humiliation complete, they invited Him to a showdown at which they would demonstrate their marvelous achievement to the entire world and then banish Him forever.

The venue was to be a wide beach by the ocean with high cliffs all around, where countless tens of thousands could witness the event. Every news network on the planet was present. The hype leading up to the showdown was unparalleled in human history.

All the world's religions had their dignitaries on hand to gloat over Jesus' inevitable degradation and ignominy. Speeches had long been prepared and slanted to the inevitable outcome. Laws outlawing every trace of

Christianity had been prepared in advance to be enacted immediately upon the success of the scientist's challenge; so vehement were the voices that put their trust in evolution and science and not in the teachings of the Bible.

It was time. An elaborate open-air laboratory had been erected not far from the water's edge. Cameras were running. Commentators were generating their own cacophony of diatribes against the church and Jesus. Godless dignitaries from every cult and false religion were reveling in what was about to happen, congratulating themselves on the long-awaited putdown of the Son of God. The audience was a sea of differing opinions and expectations.

The multitude suddenly burst into cheers; the cavalcade of scientists arrived and took their position in front of their laboratory equipment. Moments later the masses again went wild; Jesus had arrived. All was now ready. Each side took their respective positions.

The scientists, relishing in their accomplishments, smugly pontificated on how easy it had been for science to create human life using the simplest of substances and processes. The majority of the masses glowered at Jesus and hurled insults at him for how He had deceived the world into thinking that He alone deserved the title of Creator.

It was time for the crowning event: Creation!

The lead scientist outlined the process they were about to demonstrate. Jesus sat serenely across from him, totally at ease. The din from the crowd was deafening. The lynching was about to be consummated. This is what the world was waiting for, sadly reflective of a similar event in Jerusalem almost two thousand years earlier.

Two scientists took up buckets; one went to the sea and brought back seawater, and the other stepped toward the crowd and scooped up a pailful of sand. Both were dutifully placed on the table in front of the arrayed scientific technologies.

"And now, ladies and gentlemen, we create life!" a scientist shouted with great pomposity, stretching out every syllable and letting the masses go wild with each word. "All we need is sand and seawater," he further intoned in his pretentious, drawn-out way, "and we will create life from it."

Just as he began to take samples of the sand and seawater, without warning Jesus stood up, regal and ramrod straight, and walked over to him. A challenge was in the air.

The audience fell mute. Not a sound could be heard, not from the occasional gust of wind in the microphones, not from the waves, not from anything. Jesus was about to speak, and the world had suddenly become still before Him.

"Excuse Me," he said. "That's My sand and My water you're using. I made them. Make your own."

And He sat down.

Endnotes

[1] Using gallons, divide 51 million gallons (197,000,000 liters) by 23,000 gallons (87,000 liters) per tank to get the number of railway tank cars this volume would fill, then multiply by 65 feet (20 meters) per tank car to get the total length of the train, and finally divide by 5,280 feet per mile (1600 meters per kilometer) to convert to the train length to miles (kilometers).

[2] Seventy beats per minute, times 60 minutes per hour, times 24 hours per day.

[3] One hundred thousand beats per day, times 365 days per year, times 75 years.

[4] Pumping Myocytes: Beating in Unison," *Cells Alive*, http://www.cellsalive.com/myocyte.htm.

[5] The number varies depending how you do the counting. Some would count the various bones in a given structure (like the three in the inner ear) as a single bone; others would count each bone separately.

6 The pH is a measure of how acidic or basic (i.e., alkaline) a liquid or solution is. The scale runs from 0 to 14; battery acid has a pH close to 1 (very acidic), plain water at 20 degrees C has a pH of 7, and things like ammonia or lye are close to 12 or 13 (very basic). Your stomach's hydrochloric acid pH ranges between 1 and 3.

7 This is "the natural pattern of physiological and behavioral processes that are timed to a near twenty-four-hour period. These processes include sleep-wake cycles, body temperature, blood pressure, and the release of hormones. This activity is controlled by the biological clock, which is located in the...hypothalamus in human brains." "Definition of Circadian Rhythm," About.com, http://sleepdisorders.about.com/od/glossary/g/Circadia nRhythm.htm.

8 Body Worlds, www.bodyworlds.com.

9 Charles Darwin, *Origin of Species*, chapter 6, "Difficulties of the Theory, Organs of Extreme Perfection and Complication."

10 My query was "How many atoms are there in the universe?" at WolframAlpha, http://www.wolframalpha.com/input/?i=how+many+atoms+are+there+in+the+unverse.

11 Not every author has been identified by name, so setting a finite number is impossible; some who are nameless have simply been identified by their writing styles.

12 These pages were usually a thin material made from the skin of an animal (also called parchment in those days).

13 CE means "of the Common Era," a secular equivalent to AD.

14 BCE means "before the Common Era," a secular equivalent to BC.

15 For other ancient authors see Matt Slick, "Manuscript Evidence for Superior New Testament Reliability," Christian Apologetics and Research Ministry, http://carm.org/manuscript-evidence.

16 The origin of the calendar designations *BC* and *AD* are ascribed to the birth of Jesus of Nazareth.

17 A scribe could have been a copier of documents, a record-keeper, a lawyer or a judge.

18 Michel de Nostredame (aka Nostradamus) "(14 or 21 December 1503—2 July 1566), usually Latinised to *Nostradamus,* was a French apothecary and reputed seer who published collections of prophecies that have since become famous worldwide. He is best known for his book *Les Propheties* (The Prophecies), the first edition of which appeared in 1555. Since the publication of this book, which has rarely been out of print since his death, Nostradamus has attracted a following that, along with the popular press, credits him with predicting many major world events." "Nostradamus," Wikipedia, http://en.wikipedia.org/wiki/Nostradamus#cite_note-2.

19 The Bible is divided into books, chapters and verses, as in the *book chapter: verse* format used here.

20 Throughout the Bible, the words *Lord* and *God* have the same essential meaning, and you can consider them equivalent as you read on. When *Lord* refers specifically to Jesus and conveys the meaning of "Master," it has been emphasized.

21 A city in modern-day western Turkey.

[22] Pontius Pilate was the Roman governor of Judea at that time.

[23] Flavius Josephus, *Antiquities of the Jews,* book 18, chapter 3, section 3.

[24] David was the second Old Testament king of the nation of Israel.

[25] Solomon was a son of King David, as mentioned in 2 Samuel 12:24.

[26] Nathan was a son of King David, as mentioned in 2 Samuel 5:13–14.

[27] Ahab and Jezebel were the evil parents of Athaliah, who married Joram. Because of the parents' sins, the next three generations—those of Ahaziah, Joash, and Amaziah—were cursed, as per Exodus 20:5. Traditionally excluded from the geneological records because of the sins of their evil ancestors, nonetheless they are true generations in the lineage between David and Jesus and are included in the 32 generations counted.

[28] Arthur C. Custance, "Why Mary?" in *The Seed of the Woman* (Brockville, Ontario: Doorway Publications, 1980), http://www.custance.org/Library/SOTW/Part_III/chapter22.html; and ldolphin.org/2adams.html.

[29] There were two possible Bethlehems to choose from. One was a town in the area populated by the tribe of Zebulun; the other was Bethlehem Ephrathah, the ancient name for the town of Bethlehem in Judea.

[30] Justin Martyr, *Dialogue with Trypho,* 78:4-5.

[31] Publius Sulpicius Quirinius (c. 51 BCE—21 CE) was a Roman aristocrat who, as governor of Syria, carried out the census in Judea mentioned in Luke 2:4–7.

32 See map of Judea at Bible History, www.bible-history.com/geography/maps/map_palestine_judea.html.

33 An ornamental rod or staff carried by rulers or monarchs, symbolizing their supreme authority.

34 Herod the Great ruled Galilee and Perea from c. 74–4 BCE.

35 I.e., a cub.

36 Ernest L. Martin, *The Star That Astonished the World* (Portland, OR: Ask Publications, 1996).

37 Ray Bohlin, "The Magi and the Star of Bethlehem," Probe Ministries, http://www.probe.org/site/c.fdKEI MNsEoG/b.4220725/k.58F3/The_Star_of_Bethlehem.htm.

38 God had proclaimed that the seventh day of the week should always be a day of rest (Exodus 20:8–10).

39 Emmaus was an ancient town located approximately 7 miles (11 km) northwest of present-day Jerusalem.

40 *Zion* is synonymous with Jerusalem and the Jewish people.

41 Bethphage was believed to have been located on the Mount of Olives, just outside Jerusalem.

42 Twelve months of thirty days each was one calendar year for the ancient Israelites.

43 Josh McDowell, *The New Evidence That Demands a Verdict* (Nashville, TN: Thomas Nelson, 1999), 197–201.

44 Also known as the Gregorian calendar.

45 This simply means they used the money to purchase a field owned by the local potter.

46 Yosef Ben Matityahu (a.k.a. Flavius Josephus), *The Antiquities of the Jews; The Jewish War* 5.11.1 [5:446-451]

47 Ancient Persia is modern-day Iran.

48 F. Jacoby, section 256 ("Thallus") in *Fragmente der griechischen Historiker,* which superseded the previous work of Carolus Müller (*Fragmenta Historicorum Graecorum,* 1840—).

49 The chief work of Africanus was *Chronographiai* ("universal history") in five books, from the creation of the world, which he placed in 5499 BCE, to 221 CE. This work is lost, but a considerable part of it was extracted by Eusebius in his *Chronicon.* The fragments of this work are given by Gallandi (*Bibl. Pat.*) and Routh (*Reliquiae Sacrae*).

50 "Little Known Facts about Solar Eclipses," NASA, http://eclipse99.nasa.gov/pages/amazing.html.

51 Phlegon, *Origen: Contra Celsum (Against Celsus),* book 2, XXXIII, http://www.earlychristianwritings.com/text/origen162.html.

52 Tertullian lived from approximately 160 to 220 CE.

53 T. D. Barnes, *Tertullian: a Literary and Historical Study* (Oxford: Clarendon Press, 1971).

54 *Meridian blaze* means noon, at its zenith.

55 Tertullian, *Apologeticum,* http://www.earlychristian writings.com/text/tertullian01.html.

56 Most authorities identify Bethany as the present-day town of al-Eizariya, located about 1.5 miles (2 kilometers) east of Jerusalem.

57 J. N. D. Anderson, "The Resurrection of Jesus Christ," *Christianity Today,* March 29, 1968, 5–6.

58 This was in the early 1970s.

59 Charles Colson, "The Paradox of Power," *Power to Change.*

60 This approximation varies slightly depending on which biblical scholars' opinion is accepted.

61 In light of the evidence, I've capitalized the title *Creator* from this point forward.

62 In respect for Jesus' authority as Creator, from this point on I will capitalize all pronouns and titles that refer to Him.

63 Ephesus was an ancient Greek city near the present-day city of Selçuk, Izmir Province, Turkey.

64 Corinth was a city-state on the Isthmus of Corinth in Greece.

65 As linguists discovered new insights into the original meanings of ancient words, metaphors and figures of speech, and as more modern ways of saying the same thing were decided upon—like using "you" instead of "thou" or "died" instead of "slept with his fathers"—new translations appeared. These translations remain true to the original Scriptures but have been translated into more contemporary wording.